LANDSMART

'A terrific book. Every page is a challenge to how we think about the good earth about us, our climate, our soil, our food and how we live our lives. *Landsmart* is a call for positive change that we must not ignore'

Michael Morpurgo

'We all know about being climate smart, but as Tom Heap shows in this remarkable book, if we really want to save the planet we're going to have to be land smart too'

Mark Lynas, author of *Six Degrees*

'A brilliant weaving together of surprising facts and charming encounters. Tom Heap, a world class broadcaster, combines an infectious curiosity with a no-nonsense determination to get at the truth of what really works and what doesn't. Essential reading for anyone concerned about the future of energy, food and nature, which ought to be all of us'

David Shukman, former Science Editor BBC News

'Rising demand for food, energy and carbon capture all place increased pressure on land, in the process reducing that available for nature. It's a huge issue, yet how to use land in smart ways is one of the least discussed questions of our times'

Tony Juniper CBE, environmentalist

Tom Heap is a regular presenter on BBC One's *Countryfile*, specialising in the more investigative films, and has made many BBC Panorama documentaries on food, energy and the environment. Tom is also the presenter of Radio 4's *Rare Earth* series and was the anchor of *The Climate Show* on Sky News. He was the creator and presenter of BBC Radio 4's flagship climate change series *39 Ways to Save the Planet*.

LANDSMART

A Practical Guide
to Transforming
Our Countryside

TOM HEAP

Atlantic Books
London

First published in hardback in Great Britain as *Land Smart* in 2024
by Atlantic Books, an imprint of Atlantic Books Ltd.

This paperback edition published in 2025 by Atlantic Books.

10 9 8 7 6 5 4 3 2 1

A CIP catalogue record for this book is available from the British Library.

Paperback ISBN: 978-1-83895-340-9
E-book ISBN: 978-1-83895-339-3

Printed and bound in Great Britain by Clays Ltd, Elcograf S.p.A.

Atlantic Books
An imprint of Atlantic Books Ltd
Ormond House
26–27 Boswell Street
London WC1N 3JZ

www.atlantic-books.co.uk

Product safety EU representative: Authorised Rep Compliance Ltd., Ground Floor,
71 Lower Baggot Street, Dublin, D02 P593, Ireland. www.arccompliance.com

MIX
Paper | Supporting
responsible forestry
FSC® C018072

For Caroline Drummond,
who repeatedly outsmarted the prevailing wisdom
to prove that Farming and Nature
are not inevitable enemies.

CONTENTS

Introduction   1

1. The Home and Garden   17
2. Energy, Part One: Solar   43
3. Energy, Part Two: Beyond Solar   78
4. Farming, Part One: Arable   102
5. Farming, Part Two: Livestock   123
6. Peat and Carbon   150
7. Woodland   174
8. Nature   207
9. Science   230
10. Behavioural Change   263

Conclusion   282
Acknowledgements   295
Index   297

INTRODUCTION

'I've got people battering down my door, offering money to use my farm as "solar land", "battery land", "carbon storage land" and "biodiversity net gain land" but I am genuinely torn because I want to grow food.' These are the words of a Hertfordshire farmer.

It used to be so simple. Land was there to provide space for us to live and somewhere for our food to live too. Whether hunted or gathered, that nutrition needed a dwelling place. However, as time went on, food became less chased and more grown: that way our meals were more reliable. Shelter became permanent as we farmed and increasingly robust as we wandered north into colder climes. Those buildings demanded wood for walls and as fuel to heat the space. We learned how to 'grow' clothing, too, with wool, leather and cotton. As our settlements expanded, we used land to link them. And as our wealth grew we needed more land to make stuff, and as our population grew we used yet more land to feed us. We found energy below ground; at first coal, then oil and gas kept us warm, moving and powered. Many more of us lived long and prospered.

All the while, wild land – forests, meadows and wetlands – shrank, gobbled up by the plough, the cow or concrete.

The animals that dwelt there vanished and the carbon locked up in the land was released to combine with oxygen and increase carbon dioxide, CO_2, in the air. This joined forces with pollution from all those fossil-fuel furnaces, dangerously overheating our atmosphere.

That is where we are now – one in four species are facing extinction and the Earth's atmosphere is perilously close to 1.5 degrees Celsius warmer than the pre-industrial average. Fires, floods and storms are worsening and the world's fundamental geography is changing as ice shrinks from peaks and the poles. How we use land is pivotal to our success or failure in tackling these existential problems, and smart people are waking up to this.

Our use of land is a lever we can pull either way. Ibrahim Thiaw, Executive Secretary United Nations Convention to Combat Desertification, sums up the problem: 'As a finite resource and our most valuable natural asset, we can no longer afford to take land for granted. We must move to a crisis footing to address the challenge and make land the focus.'

A series of thoughtful and influential bodies in the UK have joined the chorus:

- In January 2023 the Green Alliance (an influential UK-based environmental think tank) published 'Shaping UK land use: priorities for food, nature and climate'. They argued that 'Land use must change to restore nature and achieve net zero globally. Instead of being a source of emissions, it must remove carbon from the atmosphere, while also making space for nature and food production'.

- In February 2023 the Royal Society (the UK's pre-eminent scientific academy) published a report called 'Multifunctional Landscapes: Informing a long-term vision for managing the UK's land'. It states: 'Now is a critical moment for land use policy globally, but especially in the UK. A confluence of environmental and geopolitical drivers necessitates a strategic rethink of the way decisions are made about how landscapes are managed'.
- In July 2023 the House of Lords published a report entitled 'Making the most out of England's land', arguing for a land use commission and a land use framework to help make the 'best decisions for land'.
- In November 2023 the Royal Institute of International Affairs (the UK's pre-eminent global policy think tank) published 'The emerging global crisis of land use', warning of a 'land crunch' as 'Competition for productive and ecologically valuable land, and for the resources and services it provides, is set to intensify over the coming decades, as growing demand for land for farming, climate change mitigation and other essential uses deepens.'
- The UK government's own, long delayed, land use framework is still awaited at the time of writing.

We need land to do many things for us now:

- to absorb CO_2 with trees, new marshes and managed pasture
- to grow more food for a growing population

- to provide clean energy with biofuels, solar panels and wind turbines
- to grow trees for building materials and natural fibres
- for recreation and beauty to nurture our physical and mental health
- to give space for the creatures that share our planet.

Where is all this land to come from? After all, we don't have another Earth to colonise and, with the exception for a few new holiday islands off the Gulf states, we're not creating new ground from the sea. The sad truth is that we are still stealing it from nature: since the year 2000, an area one quarter the size of Australia has been taken, the vast majority for farming and some for building.

In 2019, the first report from an international organisation called IPBES (the Intergovernmental Science-Policy Platform on Biodiversity and Ecosystem Services) was published, which marshalled research and put forward arguments to make policy makers care as much about the nature crisis as the climate crisis. Among a sackful of alarming statistics, they found that 75 per cent of the world's land surface had been significantly altered by human activity, 85 per cent of wetland area has been lost and around one million different species face extinction. The overwhelming driver of this loss has been the growth in farmland, and the chair of IPBES, Sir Bob Watson, told me in an interview: 'We must not extensify and cut down more pristine forest or [destroy] grassland or wetland'. But we still do.

In the richer world of Europe and North America, much of the wild land was cultivated decades, if not centuries, ago.

The way to fill more bellies of a booming population was to grow more food off the same area of land – a process known as intensification. This green revolution went global and, in avoiding hunger terms, was a great success as the world population grew yet famine diminished. Research from *Our World in Data* shows deaths from famine in the fifty years from 1920 to 1970 were nearly ten times higher than in the following fifty years: 70 million shrinking to 8 million. Since the global population was growing so steeply, as a proportion the drop was even higher. This was amazingly good news for humanity, but our natural environment paid a punishing price. The pursuit of productivity pushed farming to grow less variety, while using more fertiliser and the pesti-herbi-fungi-cide cocktail. In 1962, Rachel Carson's book *Silent Spring* warned of chemicals muting nature's song, with farmland becoming solely used for stock or crops and a no-go zone for anything else. It proved an accurate prophecy across much of agriculture.

But then we discovered that farming's environmental footprint is even bigger. It is a giant hose of greenhouse gases: CO_2 from cleared forest, degrading soils and manufacturing chemical fertiliser, nitrous oxide from the fertilised fields themselves and methane from rice paddies, sheep and cows. Around one quarter of human-made climate change results from farming and land use change.

Faced by the fact that farming is a major driver of both nature and climate crises, the overwhelming reaction from environmental groups has been to pressurise agriculture itself to be more wildlife- and climate-friendly. In the UK alone we have the Nature Friendly Farming Network,

LEAF (Linking Environment and Farming), Organic Farmers and Growers, the Soil Association, the Farming and Wildlife Advisory Group and many conservation groups besides. The UK's Climate Change Committee has estimated that just over 20 per cent of agricultural land must either be rewilded or converted to bioenergy or other non-agricultural crops in order to achieve net zero by 2050. All this pressure is bearing fruit: new English and Welsh government subsidies will pay farmers for the promotion of nature or other 'public goods' and not the production of food, while some landowners are opting for complete rewilding of their estates. The European Union-wide 'Common Agricultural policy' is more production focused than UK strategy but the direction of travel, both in shifting funds and strengthening rhetoric, is towards nature- and climate-friendly farming.

The business world is pushing land use in a greener direction too. Companies can now offset their greenhouse gas emissions from their transport or energy used on the production line. They are paying for the 'right' to continue polluting and make net-zero claims, by having that pollution absorbed elsewhere. These so-called 'carbon credits' often pay landowners to boost the carbon uptake of their plot by increasing soil organic matter or planting trees. There is a similar market evolving for nature with biodiversity credits. From the start of 2024 developers in England will be obliged to show that their new building project will lead to an increase in natural abundance of at least 10 per cent. This is called 'Biodiversity Net Gain' (BNG). But it doesn't have to be on the same site; damage can be offset by gains elsewhere and these BNG credits can be traded.

Campaigners, government and commerce are broadly agreed on the menu: hungry for bees, bats and birch trees but with less appetite for food. Even the war in Ukraine, triggering food security fears, hasn't fundamentally weakened this strategy. But there is an inconvenient truth: traditionally, land that delivers more abundant wildlife will yield less food, yet the world needs more food, not less, as we are expecting about two and a half billion extra mouths to feed by 2050. So where will the new food come from, especially as similar farming policies are gaining traction across Europe and North America? The answer could be an accelerated worldwide land grab from the natural world. Our well-intentioned green farming policies could increase the destruction of wild habitats, loss of species and rise of carbon emissions elsewhere. Other countries' 'Edens' may become farms so we can keep our bellies full and our local wildlife off life support. I recently asked one of the architects of England's farming and nature strategy if they examined the effect of any policy change on land use overseas? 'Errr, no,' came the reply.

The map on the next page shows the UK split according to how (not where) its land area is used.* By far the biggest chunk is related to animals as pasture or crops for feed. The hexagonal islands to the right show the amount of land needed to grow the food the UK imports. It is roughly equivalent in size to another UK.

* With thanks to the UK National Food Strategy 2021 but apologies to Orcadians who seem to have lost their island status!

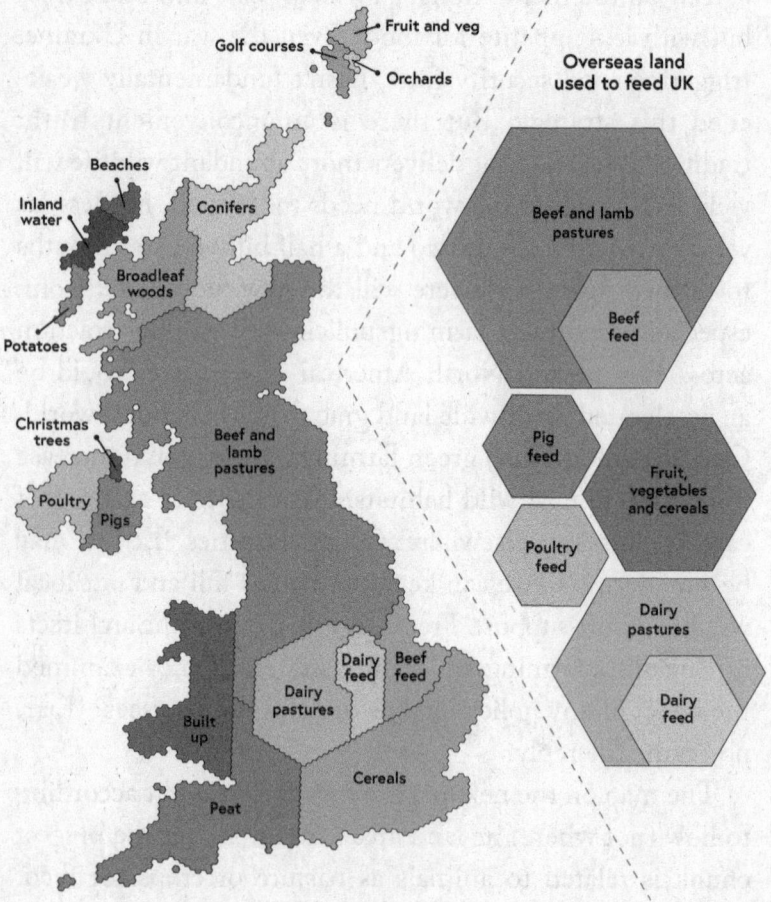

Fruit and veg

Golf courses

Orchards

Overseas land
used to feed UK

Beaches

Inland
water

Conifers

Broadleaf
woods

Potatoes

Beef and lamb
pastures

Beef
feed

Christmas
trees

Beef and
lamb
pastures

Pig
feed

Fruit,
vegetables
and cereals

Poultry

Pigs

Poultry
feed

Dairy
pastures

Dairy
feed
Beef
feed

Dairy
pastures

Dairy
feed

Built
up

Cereals

Peat

Current land use policies will enlarge this offshore island and grow the expectation that somewhere else and someone else will grow our food. Similar pressures and policies are strengthening across much of the richer world. It's no wonder that every day since 2010 we destroyed on average more than 20,000 hectares of tropical forest – an area about half the size of the Isle of Wight or twice the area of Paris.

We are faced with two ethical paths – duties, if you prefer – that appear to be in conflict: either 'we should restore *our own* natural habitats and do what is in our power to halt climate change by becoming net zero' or 'existing farmland should grow as much food as possible to avoid promoting the *global* biodiversity and climate crises.' Both of these are sound environmental arguments, but the second is one that many environmentalists still reject. However, high yield from each field should be welcomed not scorned. We have a finite amount of land on Earth and other creatures share it, so confining food production to as small an area as possible is good. Intensive farming – the enemy of green campaigners for years – needs to become an ally.

Environmentalists' hostility and suspicion of intensive farming have plenty of justification, though, as traditional intensive agriculture has a lengthy charge sheet. Firstly, in the field itself, it can lead to impoverished soil, eliminated wildlife and greenhouse gas emissions. Secondly, there is a wider footprint of water pollution, air pollution, fossil-fuelled fertiliser production and, in livestock systems, tonnes of imported animal feed and exported dung. In fact, given these broader effects, it's reasonable to suggest that far from being *intensive* (as in concentrated in one place) traditional intensive farming has an *extensive* impact overall, tarnishing much of the world.

However, this isn't just a question of farming vs nature. There are many other demands on land – green energy production, carbon storage, business parks, housing, commercial forestry, flood alleviation and leisure – and they all want a piece of the world's dirt pie. Luckily, they don't all have to be separate. In many cases, with the right management, land can do more than one thing at once: the space around wind farms can grow food and store carbon; forests can provide habitat, commercial timber and recreation; homes and warehouses can be roofed with solar; farmland can fight climate change and yield a rich harvest. Elsewhere a laser-like focus on one activity might be preferable, such as dedicating an area to the survival of an endangered species or reaping bumper crops from rich soil.

This plays into a lively debate that has taken place in farming and environment circles in recent years. Which approach to land use is the best for humanity and nature: sharing or sparing? A typical 'sharer' might be an organic farmer with more wildlife per hectare but who yields less food. A 'sparer', on the other hand, would balance maximising food production or energy generation in one area but demand total protection for nature elsewhere. I don't think either strategy has the monopoly of virtue or villainy.

So now we need to think about this: how do we get the most out of every scrap of land without the damaging side effects of pollution or expansion? Intense, huh?

The answer lies with the best farmers, land managers, scientists and conservationists who are working today and those yet to come. When it comes to their land, the people in the pages of this book share a common raw material: intelligence. They have all thought deeply about the land crisis facing the world and come up with a solution.

But there are other critical players in this game: us. The consumers of food, energy and living space make choices that are then written in the land. And, given that farming is still the world's dominant land use, food and diet is where we can make the biggest difference. Roughly one third of food grown is never eaten but wasted in the field, supply chain and home. If we cut waste we cut pressure on the land. We could also just eat less. More of the world's population eat too much than eat too little. Obesity, and the health problems often worsened by it, is affecting a growing proportion of the population. We eat on average nearly 3000 calories per day whereas the requirement (an average of men and women) is 2250. If we all ate the recommended daily calorie intake, we would reduce food consumption by one quarter. Eating less meat would ease the land squeeze too. Animals aren't very efficient in converting the calories and nutrients from what they graze into the flesh on their bones.

It is plausible that if we reduced waste to just 10 per cent of food grown, reduced overeating and lowered meat eating by half, the fall in demand for agricultural produce would leave sufficient space for nature and carbon storage across the world. But recommending behavioural change and delivering it are worlds apart.

A real stumbling block for so many environmental narratives in books, media articles or academic papers is that gap between prediction and reader experience. All the authors, in various ways, are saying: 'Things are getting worse because of our abuse of nature and we risk going to hell.' But the problem is that the bulk of the audience are not feeling even close to hell yet. Sure, we have observed the decline in wildlife and the temperature graphs rising, but this isn't enough; more people are living longer, healthier and wealthier lives (although in recent years the income inequality gap has been rising). We continue to do so well from our exploitation of nature that convincing us to stop is really hard.

Relying on behavioural change betrays a common weakness of environmental advocacy: dreaming of a world where people think and act like you, rather than finding answers in tune with real and current popular motivations. Hoping that future populations will want different things has a place, but it is no panacea.

Climate change, nature depletion and food supply are quite obviously international issues, and land use is at the heart of all of them. So should this book have a worldwide scope? There is no absolutely right answer to this and my solution is, I hope, an elegant fudge: land use examples principally from the UK that acknowledge their international context and could be applied elsewhere. The arguments for 'going global' in this book are to do with the planetary scale of the issues, the fact that decisions made in the UK have repercussions elsewhere in an international market, and the choices made in other countries – especially the massively populated India and China – are probably more impactful than those made in the UK. But there are big drawbacks to this approach. Firstly, it would be a massive undertaking that would risk either superficiality or absurd size. Secondly, there is a strong risk of hubris – always there to an extent with writers or journalists – but I simply don't have the same experience of the land issues in India or Peru as I do of those options closer to home.

Yet one must recognise the planetary scale of the problem. The World Resources Institute (WRI) is a global research organisation dedicated to delivering practical solutions to improve people's lives and ensure nature can thrive. In recent years, they have been preoccupied by this land crisis. They report that a little over two thirds of land surface is or has been used for some productive purpose: mainly grazing, cropland, forestry and the built environment. The remainder – truly wild space, if you like – is divided between vegetated natural ecosystems and the area that is barren sand, rock or ice, which is just over one tenth of the surface. The need for food has driven this land grab, with croplands and pasture now taking around half of all the vegetated area today. The expansion is continuing and the WRI estimates that between 1962 and 2010 almost 5 million km^2 of forests and woody savannas were cleared around the world for agriculture – equivalent to about two thirds the size of Australia. And here's another verse from the WRI songbook: since the start of the twenty-first century, 1 million km^2 – an area the size of Egypt – has been converted to cropland, some taken from pasture, some from previously virgin land. So more land is being taken for farming whether it's under the cow or the plough.

Pretty much the worst thing you can do for nature is enslave it to our demands, in other words: farming. Its bounty shrivels.

The number and variety of life forms plummet. IPBES says one million species are now threatened with extinction, some within decades, and that is more than ever before in human history. The primary cause is change in land and sea use; the secondary is 'direct exploitation of organisms' (such as fishing or hunting, both of which serve our appetite for food). The IPBES chair, at the time, Sir Robert Watson summed up why this matters: 'The overwhelming evidence of the IPBES Global Assessment presents an ominous picture. The health of ecosystems on which we and all other species depend is deteriorating more rapidly than ever. We are eroding the very foundations of our economies, livelihoods, food security, health and quality of life worldwide.'

Boom.

And then we get to climate. With very few exceptions, changing land from a wilderness to a farm releases CO_2 into the atmosphere. Most obviously this happens during deforestation, where the trees themselves are burned and the great majority of their stored carbon goes up in smoke. Then you have the disturbance of the soil, accelerating the rotting of plant material through greater access to oxygen, which again pumps up the amount of CO_2 released. In a typical forest 44 per cent of the carbon is in the plants, roughly the same in the soil and the remainder in dead wood and leaf litter. All of this is vulnerable and much will vanish once the land is cleared for agriculture. The farming frontier also advances into wetlands and, as we'll see in the chapter discussing peat, this also releases huge amounts of carbon.

Land use change within farming accelerates carbon emissions too. The soil beneath pasture holds more carbon on average than that beneath a crop, yet across the world in the first two decades of the twenty-first century about 1 million km² of pasture and abandoned agricultural land was transformed into arable fields – an area about twice the size of France. Overall, according to the United Nations Intergovernmental Panel on Climate Change (IPCC), land use change contributes 14.5 per cent of greenhouse gas emissions, whereas the actual practice of farming activity accounts for 'only' 8.5 per cent. Leaving wild land alone is one of our biggest climate change solutions.

Let's roll out the map again because gobbling up fresh land for food is not seen everywhere. According to Global Forest Watch, deforestation and farmland expansion are most rapid in South America and particularly Africa. In North America tree cover is roughly stable, whereas in many parts of Asia, Eastern and Northern Europe, including the UK, it is actually increasing. It should be remembered that replanted or regrown forest does not have the same natural richness as virgin woodland and takes time to build up similar carbon storage. Overall, in the first twenty years of this century, we lost 2.5 per cent of our forest cover.

So how do we halt our continuing invasion of the natural world? The WRI has delivered some robust and shocking studies on how, in the competition for land, nature continues to lose. Its solutions rest on four pillars, paraphrased here:

- **Produce:** sustainably produce more food, animal feed and fibre from existing farmland
- **Protect:** shield the remaining natural world from exploitation

- **Reduce:** produce less waste and consume less meat especially in rich countries
- **Restore:** give poor agricultural land and degraded ecosystems back to nature.

I think this is a pretty admirable list but I would add one more:

- **Combine:** where possible and without big side effects, get your land to multi-task.

There are so many jobs for soil out there. The chapters that follow are divided into the different spatial demands such as farming, energy, forest, nature – but, given that many of the solutions involve combining land uses in the same space, there is obviously some overlap.

I am not trying to compete with the grand academic reports mentioned earlier in the introduction with modelling analysis and policy recommendations. I want to investigate the real-world pressures on our planet and take you on a guided tour through the UK to meet the thoughtful innovators in fields, labs and even parking bays who are using space smartly to avoid the land crunch. So let's go and see who's putting our land to good use.

1

THE HOME AND GARDEN

Smart use of the world's land is a head-spinningly massive project, so let's start small – very small. The area I look after – my house and garden including the allotment – is probably a little shy of 1000 m², or one tenth of a hectare. The world's land area is 13 billion hectares, so clearly the direct impact of my choices will be modest, but the thought process and trade-offs that go into my decisions are relevant to the wider world, and everyone's personal plot adds up to a sizeable footprint. The amount of land given over to gardens in the UK comes to about 433,000 hectares: roughly the size of Somerset or one fifth of Wales. Within different settlements, the proportion of land that is garden space varies; in Edinburgh about one third is garden, whereas in Leicester it's close to half.

I live in a four-bedroom cottage in Warwickshire with a surrounding garden about the size of a tennis court and a separate building consisting of a workshop, guest bedroom, bathroom and home office. The allotment is just 150 metres away on a southwest-facing slope and has clay soil. So what is my patch delivering both for me and the wider world?

Shelter

I sleep, eat and mooch about here with my wife, Tammany, and (until recently) three sons. Not being in a block of flats, I am not stacking occupants for maximum dwelling space per square metre.

Recreation

The house and garden provide space for exercise and relaxation for both mind and body. Tammany is a skilled gardener whereas I'm what footballer Eric Cantona called 'a water carrier' – a necessary but somewhat dull-witted stalwart. This is literally true in my case as we are lucky enough to have a well on the property, which avoids the use of tap water in the garden but involves many watering cans.

Nature

We make a deliberate effort to encourage wildlife. The south side of the plot is fringed by bramble thicket and nettle beds, while the roadside fence wears a hefty crown of ivy at one end and runs alongside a dense hawthorn bush at the other. These provide shelter for birds' nests, as do the somewhat rotten eaves, and both teem with starlings. Most of the plants are native so local creatures, be they birds or bees, are more likely to find them a good food source. We have two tiny wildflower patches, each a little bigger than a double bed, that hum with insects in the summer. (I always think the term wildflowers is a total misnomer. It implies less human intervention, whereas no other part of the garden requires as much time per square metre. Tammany patrols regularly to be sure those fragile meadow beauties aren't bullied out of existence by tougher invaders.)

Industrially produced chemicals are largely absent. The small lawns get a fertiliser hit from so-called 'nettle tea': stingers left in a tub of water for at least three weeks to produce a horrifically stinky organic fertiliser. An allotment neighbour refers to it as 'Tom's evil brew'. Thankfully the smell fades after a few hours and the growth promotion lasts for months. The hardest wildlife to embrace are the slugs and snails as their appetite for lovingly nurtured seedlings almost has me reaching for the chemical weapons. However, those blue pellets are harmful to birds too, so our assault on garden molluscs is limited to throwing them as far as possible, beer traps and the occasional squishing.

We have a pond in the shape of a semi-circular trench about 4 metres long and 60 centimetres deep that is rich in plant and bug life. We'd love it to become a home to frogs or newts. A friend in London has a thriving colony of newts and such great pond lighting (he is a cameraman) that the amphibians' nightly forays play out like a shadow ballet cast on the water surface. We scoured the local puddles for newts and looked for frog spawn but our part of Warwickshire appeared to be amphibian-free. Thankfully, a ditch in Hampshire provided the goods and a few weeks later we had some froglets emerging from our pond. Tammany and I fondly watched the growth of our new brood (did I mention the boys had left home?) and imagined the crazy frog antics to come. The truth was a rather different wildlife triumph: a grass snake appeared, which looked beautiful, and returned regularly to our 'all-you-can-eat buffet' until it had scoffed the lot. Or so we thought. But a year later some more tadpoles have appeared. It is impossible to say if these were the offspring of snake dodgers or a new population but, either way, they are very welcome.

Energy

In an effort to cut carbon emissions from our house and a willingness to see more clean energy on the grid, we have installed enough solar panels on the roof over the years to generate nearly 8 kW (kilowatts) and a battery to store 8 kWh (kilowatt-hours). We'll look at this in further detail in chapter 2, but it has changed our behaviour, as we now try to wash dishes and clothes when the sun shines and we have reduced our electricity bill and generated income from generous tariffs. As a shared land use, in my experience, it is a complete winner.

I wish I could harness the wind. We live on a hill that faces south with a fetch in the direction of the prevailing wind of more than 8 km (5 miles). I listen to gales at night and think of all that energy whistling round the eaves and buffeting the flower-beds. I want some of that feeding my battery, heating my water tank or warming my walls. Small domestic wind turbines had a brief flurry in the early 2000s with even the then prime minister, David Cameron, fitting one on his house in London. But they were over-hyped and underperforming. Turbulent city winds delivered a fraction of the claimed potential wattage and losses were amplified by the inverters required to turn this wildly varying direct current (DC), which is delivered by a turbine, into stable useable alternating current (AC) that flows from your sockets. The results were generally so poor that it came close to a mis-selling scandal and certainly tarnished the reputation of household turbines. Fifteen years on there are only around 125 domestic turbines installed across the UK (compared to around half a million domestic solar arrays). Some basic problems remain: solar panels with a similar power output are cheaper; turbines are not expected to last as long as photovoltaic (PV) systems as moving parts don't help longevity; they make some noise; and I would expect the neighbours to be more concerned about the sight of them than the existing solar panels.

Food

Homegrown food has been part of my diet all my life. Where I grew up, near Cambridge, we had a big garden with a vegetable patch and the gnarled remnants of an orchard: mainly a popular variety of cooking apple called Bramley. There was one 'Beauty of Bath', which gave a poor yield but excellent support for a rambling rose. The trees were my climbing frame throughout the year and in autumn I loved the challenge of reaching the biggest, highest fruit before lobbing it down to my parents below. We always had far too many to use at home despite baked apples and crumble being a staple dessert. Some were stored in an old concrete air raid shelter, where massive walls ensured stable temperatures; others were offered free at the gate, while the remainder fed hungry bees, drowsy wasps and the soil. As a boy I was less excited by the vegetable plot – it lacked things to climb – but I remember good yields of potatoes and beans. I lived in London for about twenty years and for most of that time found space for growbags and a few window-ledge seedlings.

Now, in Warwickshire, the garden round the house is mainly planted for beauty, nature and leisure: flowers, shrubs and grass. But a few herbs and salad leaves find a space. Then there is the greenhouse: my pride and joy. Built from stone and timber by my family and I, it yields tomatoes and chillies while leaking so badly it has gutters on the inside. The produce is delightful and keeps me in tomatoes for months while the chutneys and chilli sauces keep things zesty all year round.

It is the allotment, of course, where food is the priority. About one third of it is dedicated to different potato varieties. One called 'Pink Fir Apple' is a particularly tasty favourite, while 'Charlotte' are grown for heavier cropping. We also grow sweetcorn, beetroot, onions, courgettes, spinach, squash, peas, salad greens, globe artichokes and the obligatory runner beans. Raspberries and blackcurrants deliver the sweet course. I've never done the research to measure the yield but I would guess that, across the year, we grow about one quarter of the fresh fruit and veg we eat. I love the quality and satisfaction of 'growing your own'; the food nourishes my body and the labour is some defence against indolence and heart disease. It is probably cheaper than the shops, so long as you don't charge for your own hours.

Carbon

This is my recent obsession, inspired by farmers who are turning their fields from carbon sources to carbon sinks. Apart from what is eaten and the really pernicious weeds, everything that grows from our garden and allotment goes back onto the soil: grass cuttings, tree trimmings, old vegetables, wilting shrubs, ivy and sawdust. Kitchen waste has its own compost bin but it all has the same eventual purpose: delivering more food and capturing more carbon. I ferry it all to big boxes on the edge of the allotment in the same green bins that were once emptied by the council and there it sits for about year. In the early spring we empty the boxes and spread the contents over about one quarter of the ground before roughly chopping it up with a spade and covering it entirely with a woven plastic weed-blocking membrane. There it sits for another year, hopefully boosting both the fertility and carbon content of the soil. I now can't think why I ever let the council take this valuable raw material away, especially as in many places now you have to pay for the service.

Another way to keep more carbon in the soil and less in the atmosphere is to dig less: the gardener's equivalent of limiting ploughing. When the top layer of soil, which is full of carbon-rich organic matter, is disturbed by the spade or the ploughshare and exposed to oxygen it causes oxidation of organic matter within the soil, which releases CO_2 into the atmosphere. Worms and fungi in the ground also prefer not to be sliced by steel. This presents me with a challenge, though: how to deal with compacted clay soil and persistent weeds. The spade and the rotavator have not yet been retired but they have gone part time on very limited hours. It's a faint echo of what many farmers encounter in trying to make this change.

When it comes to land use, our house, garden and allotment offer a multifunctional space with food, shelter, energy, nature, carbon and recreation all given room. I think we are using our 1000 m² well, by which I mean we're not wasting the area and it is delivering a good chunk of what we need land on this planet to do. I accept that we have more room than many and more money than some, but I think there is some lesson here about engagement: my personal patch is delivering a lot because Tammany and I think about it, we care about it, we enjoy it. It is intensively managed.

However, compared to some, we are just scratching the surface. Meet urban gardening maestro Mark Ridsdill Smith: 'From an area of about two by six metres, I grew 90 kilos of food worth about £900 and ate something fresh from it pretty much every day of the year.' Mark didn't even have a garden, so this was all from pots and tubs on a balcony in north London. He grew mainly herbs, tomatoes, chillies and salad veg, with some runner beans and potatoes in season: 'It was about adding value and diversity to my diet rather than big calories. But these aren't just frills for foodies. They are essential nutrients.'

Mark is the creator of Vertical Veg: a website, forum and book all about container gardening. It gives those without land the power to grow food. Mark recently had his book on vertical veg published in famously foodie France. (The French title is *Y'a des legumes au balcon!* – *There are vegetables on the balcony!* The tone of alarm suggested by the exclamation mark makes me chuckle: 'There are zombies on the balcony!')

Mark began making a career from a passion when he was made redundant from the charity sector in 2009. Living in a flat in north London with no garden but some time on his hands, his first thought was an allotment, but the waiting list for a plot in Camden was close to thirty years. Disappointed but not discouraged, his ambition sprouted on windowsills and a small flat roof: 'I had a real desire just to grow food. It's such a pleasure. I think of homegrown food as 3D food, not the one-dimensional stuff you get processed or wrapped in the supermarket. And it uses all the senses: the smell of earth and tomato leaves, the sound of bees or a popping broad bean pod, feeling that pod's furry lining. I think this is something very deep within us. We have been growing and harvesting for most of our history, the disconnection of shops is only very recent.'

Strangely, he doesn't immediately mention eating it: 'The stronger flavours and greater variety of what I grow is great. But eating wasn't the main reason why I did it in the first place; it was about the process and experience of growing.' With food, as with much else, the journey enhances the destination.

Mark now lives in a suburb of Newcastle upon Tyne and has a garden. He invited me to see his new plot and, when we step out of his back door, I expect be greeted by a food forest. But no: there is an overgrown lawn surrounded by an abundant flower-bed. The front yard, though, which is entirely paved, is a vegetable Eden. 'What's going on?' I wonder. Does he feel actual earth is too easy? Does he have some kind of horticultural Stockholm syndrome where he's compelled to behave as if still confined to a flat? 'It's a deal with my wife, Helen. She didn't want to live in a veg patch so the lawn and surrounding flower-beds are her domain.'

As with my home, compromise delivers shared land use: beauty and recreation alongside food and shelter. But let's get back to that slabbed front patio. It's just about big enough to park a car and open the doors to get the grocery shopping out, and doubtless it once did just that. But no longer. Everything is in containers and Mark gives me a potted botanical tour:

'Coriander. Incidentally grown on from live herb pots from the supermarket. Purple raspberries, orach (which is a kind of purple spinach), rhubarb, Scots lovage, blackberry, blueberries, sorrel, peas, strawberries, lettuce, courgette, rocket, runner beans, fat baby achocha, mint, society garlic, chocolate mint, chives, marjoram, oregano, Japanese wineberry, blackcurrant sage, garlic chives, parsley, thyme, plum, three-cornered leek, cherry fennel, apple tree, one I can't remember, radishes, Chilean guava, nasturtium, sunflowers, pea shoots, hardy ginger, Jerusalem artichoke. But it faces north, so I do have some space on the south side for sun-lovers like chillies, rosemary, tomatoes, mulberry, sage, loganberry, lemon verbena, tarragon, savoury, lemon thyme and a hardy kiwi that has never produced fruit but it has lovely flowers and I remain hopeful.'

That is nearly fifty foodstuffs and there is still room for food for pollinators like aquilegia, allium, cosmos: all growing without direct access to the earth. Stunning. His front yard is a head-turning cornucopia of fresh fruit and veg, making the rest of his street seem rather barren. But he can't live on it as, with a few exceptions, these are not staple crops that deliver the baseload of protein and calories. Mark and Helen have two children and he reckons the garden is providing less than 5 per cent of their calories across the year: 'I was thinking about focusing on growing calories just for an experiment. I'd grow potatoes, Jerusalem artichokes, runner beans, tomatoes and keep the fruit trees. But it would be a pretty boring diet.'

Mark is not trying to make a fundamentalist point about being self-sufficient and he says that farmers are best placed to grow staple crops. He is growing for joy, nutrition, variety and health. And, because much of what he harvests at home is expensive in the shops, he reckons he's shaving around 20 per cent off the weekly food bill.

I visit Mark's home garden in May and nearly everything is looking healthy and verdant. His front door is becoming obscured by foliage but I wonder where all this fertility is coming from. I suspect Mark is not the type to reach for the chemical fertiliser: 'The main thing is worm compost. Let me show you these two wormeries.'

Mark lifts the lid on two homemade wooden crates, revealing a surface of rotting green matter that quickly transforms into friable brown compost as he digs down. Worms trickle through his fingers with every fistful: 'All our food waste goes in the top, along with prunings from the vegetable garden and occasional helpings of nettles, comfrey or manure. That is the main engine of the garden and it delivers so much to the plants: nutrients, minerals, but the really brilliant thing is the microbial life, which activates the soil and makes food available to the plants.'

Elsewhere buckets full of nettle or comfrey 'tea' are brewing away to make pungent liquid fertiliser. Mark then lowers his voice and checks who's in earshot: 'More and more I'm trying to make homemade things. Helen doesn't know about it but I'm trying a new brew, which is fish and sugar. She might think it a bit too committed. I get fish scraps free from the market, and when you add the sugar and a little water it ferments to a nice-smelling liquid root drenching or foliar feed.'

When it comes to helping his plants grow, he has a hierarchy: garden waste, kitchen waste, somebody else's waste and, only when the waste streams are exhausted, organic fertiliser. I say 'exhausted' but Mark says there is a big waste product he is not using: what four humans flush down the loo. The property and lifestyle don't permit a compost toilet option, but he would if he could. (In chapter 9 we'll look at the enormous, yet under-used, potential of the sewage system to help our plants grow.)

Aside from growing food, Mark's other passion is spreading the word about growing food, hence the website and book. He also ran projects in the neighbourhood where they went door-to-door and offered the occupants free microgreen seeds and a tray. Of those who answered the door in the most recent campaign, 80 per cent said yes, which Mark thinks is a pretty good return from cold-calling. Most of them had grown something two weeks later, which he hopes will provide them with the encouragement to try something harder. He called the project 'Random Acts of Greenness'.

Random door-knocking is one way to overcome the emerging middle-class bias in homegrown food. Mark has seen it at every stall he runs, at every talk he gives, even in the gentrification of the allotments, and considers it an especially unwelcome trend. It is usually poorer urban areas that contain 'food deserts', in which accessing healthy fresh food is very difficult, and therefore are places where people would benefit the most from growing their own. The socio-economics of our food culture deserves a book of its own, but Mark does acknowledge some real barriers for those on lower incomes: 'Gardening goods tend to be heavy, you may not have a car, and there are few role models, but the biggest impediment is lack of knowledge and confidence. If you have managed to scrape together £20 one month, are you really going to risk it on growing something that may die and not feed anyone? Freezer food or takeaways are reliable.'

Yet it remains perplexing to me and it is a relatively new phenomenon. A study of the relative productivity of farms, garden and allotments produced in the 1950s suggests that, back then, it was those on lower incomes who were much more likely to grow veg. The study compared the use of gardens in suburban London and found food production in 21 per cent of the council house gardens and only 9 per cent of the private homes. It is a fair assumption that poorer people lived in the council homes. Council properties also had a higher proportion of derelict gardens, whereas the private holdings had 50 per cent more lawns, flowers and shrubs. To me the obvious reading of these stats is that not many hard-up households could afford to waste their time gardening for decoration; their tilling hours had to yield something to eat. I wonder why that has changed.

As I leave, Mark warns me not to step on what I had assumed was a weed, growing out of a crack in the tarmac beside his gate: 'That is one of my salad leaves, Orach Scarlet Emperor. Elsewhere around here, I've spotted self-seeded chives, marigold, society garlic and verbena persevering through the cracks in the pavement. You can't stop the seeds.'

Mark Ridsdill Smith is a powerful and practical prophet for urban gardening, but to what extent can individual or neighbourhood enthusiasm really fill the nation's veg box?

In the outwardly affluent area of Oxfordshire, Dr Emily Connally founded Cherwell Collective, a non-profit organisation that works to reduce climate change and social inequality through growing food and tackling waste. She says she is growing food to meet demand from people on lower incomes: 'With the help of food bank users and volunteers, we bring redundant spaces into productive gardens. We now look after 5–10 hectares in total in the outskirts of Oxford itself and nearby towns of Bicester and Banbury. We distribute hundreds of meals and a few tonnes of food every week.'

Dr Connally has a passion for forest gardening. The idea is that you try to mimic the structure of a woodland but with plants that yield food. When it comes to maximising photosynthesis off a given area, woods do it best and so, the logic goes, you should do that but with food. From top to bottom, this means fruit trees, vining or climbing peas, beans or squash to twist up them, fruit-bearing bushes, leafy greens, onions or leeks and then root crops like carrots, beetroot and potatoes. Herbs and even mushrooms can find their place too. And she is obsessed with putting all the dead stuff back on the soil: 'Why does society bag up and throw away leaves and then buy compost? You are actually buying leaves and old branches. Keep your own and have a little patience.'

She grew up in the desert of New Mexico in America, where there were no nutrients in the soil and it only rained for two weeks a year. It taught her to value moisture and organic matter – principles that still serve her well in the relatively fertile Cherwell Valley. She's been in the UK for twelve years and running Cherwell Collective for three. In that time the gardens have been getting more productive as they mature and, when we speak, it's the middle of the early-autumn glut with a tonne and a half of surplus produce from their own gardens and tonnes more donated. The immediate emphasis is on pickling, chutneys, soups and freezing both fresh produce and whole meals. These skills are taught and spread, not hoarded in a hub.

Dr Connally believes urban gardening can really help with the staples too: 'We have worked out that a 10 metre by 10 metre plot can keep a family fed over the twenty-week

winter, from November through February. Plant a combination of broccoli, sprouts, chard, kale, turnips, beetroot,
parsnips, winter salads and parsley.' Her motivation is spelled
out on the website: 'We empower those in our community who, due to social, financial, or medical inequities and
exclusions, believe that reducing our impact on the climate
is beyond their reach.' Though waste avoidance and environmental concern are her major drivers, she cuts that out
in most communications with vulnerable groups, focusing
instead on the cost savings and health benefits because these
are more likely to cut through as immediate concerns. But
the idea of using gardens to help the underprivileged and
enhance our cityscapes goes well beyond Oxfordshire.

Spread out on the floor beside my desk are maps
of Newcastle upon Tyne, Hackney (east London),
Peterborough, Brighton, Bishop Auckland (County
Durham), Vickery Meadow (a suburb of Dallas) and
Domiz refugee camp in northern Iraq. They are maps of
the potential for urban gardens including rooftops, park
edges, abandoned land, indoor farms. They were made by
urban farming zealot and guerrilla map-maker Dr Mikey
Tomkins: 'People say there is not the land available and
people say you are not going to grow a lot of food. There is
the land available – you have to map it, you have to collect
the data. It doesn't have to be "food vs existing land use", it
can be an inclusive conversation to include the multifunctional nature of space.'

These 'edible maps' appeal to my taste as a geographer
who also loves food. They are maps of a possible future
where our living and working realms are combined with

growing space. Potential plots in each settlement are high-
lighted with a similar key:

- **Rooftop gardens:** top decks of carparks, flat roofs on
  retail or office space and our own homes
- **Meanwhile gardens:** spaces awaiting development
  (this can be many years) turned over to growing
  food or flowers and reducing urban blight in the
  'meanwhile'
- **Tarmac gardens:** raised beds and containers on closed
  roads and excessive hard-standing
- **Indoor gardens:** empty buildings used to grow crops
  under lights
- **Accidental agriculture:** food that grows wild in the
  city like herbs, plants for pollinators and trees.

Once fully developed, this would become what Dr
Tomkins calls 'Continuously Productive Urban Landscape'
– a network of space and green infrastructure that includes
urban farms, orchards and allotments, for food biodiversity and relaxation. It assumes that cities of the near future
will need less space for cars and he is particularly animated
about the wasted potential of the top deck of multi-storey
carparks: 'People don't go up there. It's further to walk to
and from your car. The vehicle gets too hot in the summer
and you might have to dodge the elements in the winter.
They are very accessible and structurally sound spaces. Very
multifunctional so perfect for rooftop growing.'

As I write, there are proposals for 'Slow Food Birmingham'
to take over the top floor of a carpark in the city's Jewellery

Quarter. In Paris, a roof space of a little under a hectare is now covered with vertical growing tubes sprouting salad leaves and herbs along their 'trunk'. In Lisbon, local authorities have earmarked temporarily available space for mushrooms in shipping containers growing on waste coffee grounds. There are plenty of other veggie plots across many cities but, with a handful of exceptions to be explored later, they tend to be niche. Why? Dr Mikey Tomkins believes that: 'The food system, though creaking, is still delivering. And the gap between simply going shopping and becoming a food gardener is so vast that it will take a crisis for most people to get into it.'

There is no denying the hours of labour needed to grow food on a small scale compared to a massive farm where cutting human input has been ruthlessly pursued in recent years. On arable farms machines rule, while even in sectors like salads or soft fruit that still need dextrous hands, the human element in the system is far smaller than in home growing. But it's not only the quantity of work needed to 'grow your own' that is off-putting for many but also the type of work. Mikey says: 'Some studies in South Africa suggested recently that people didn't want to get back into growing their own food as that is something they *had* to do under apartheid. "Why are you trying to keep us as farmers?" they asked. People have an association with urban agriculture as something you do under crisis: war, poverty, oppression. Not something you do voluntarily. Put even more simply, urban food growing is something we do when we *have* to do it.'

But when we had to do it, the figures suggest gardens can be remarkably productive. The most well-known historical role model for ambitious urban farming in Britain is the wartime 'Dig for Victory' campaign. Before the Second World War Britain had relied very heavily on imports, with 70 per cent of food coming from overseas, much of it from existing or former colonies. But as the German submarines – U-boats – patrolled the oceans and torpedoed incoming grain shipments, there was a massive push for domestic production, both on farms and quite literally at home. Domestic gardens switched from lawns and flower-beds to veg patches. Even city parks were ploughed up. 'Dig for Victory' became a persuasive and popular slogan and, given that starvation was one of the causes of German defeat, not without some truth.

A 1956 academic study by 'Best and Ward' trawled the archives for data on productivity during the Second World War in the UK. It delivered one million tonnes of extra food, with food imports cut by half. Studies also suggest that the yield per hectare (or square yard back then) was higher in cities than on farms. By 1944, according to a parliamentary reply, the produce from gardens, allotments and similar plots of land represented 10 per cent of home-produced food. For pork it was approaching 15 per cent and for eggs close to 50 per cent; pigs and chickens are both potential backyard animals partial to waste. Given the relatively small area given over to food production in built-up areas vs the countryside, these are pretty staggering figures. The 'Best and Ward' paper also includes a table following research in 1941 on the relative productivity of allotments and farms for different vegetables like potatoes, leeks, beetroot and peas. The yield per unit area is greater for allotments than farms:

17.5 tonnes per hectare from the allotment compared to 15.5 tonnes from the same area on the farm.

The authors acknowledge that weight is a pretty blunt metric, as a kilo of beef has much more protein and calories than a kilo of lettuce. Elsewhere they quote studies that look at the monetary value of food from the two systems – and once again domestic production seems to trump farming. The precise figures are of limited use as they are decades old and both farming practices and food prices have shifted massively. But these and other more recent studies suggest that food production from urban areas is low because we choose it to be, not because it has to be. Or, to put it another way, wartime experience reveals what we *can* do but the question is what we *want* to do.

As an interesting aside, the 'Best and Ward' paper was being written at a time of great urban expansion when the suburbs of London and other British cities were sprawling over the surrounding farmland, yet the memory of wartime food shortages was still fresh. Some said this loss of productive fields could make us vulnerable once again, while others argued that the new suburbs with reasonably sized gardens could produce as much food as the farms they replaced (assuming, of course, that the new residents were very keen growers). In the event, many new housing estates were built and the lengthening distance from the war weakened the hunger for food from the home front.

Yet Dr Mikey Tomkins and many others believe we should rediscover that appetite because – whether measured by farming's massive harm to nature, a tight global food system exposed to shocks like the war in Ukraine, or our processed diet worsening our health – we are in a food crisis. He says urban agriculture doesn't propose self-sufficiency, just 15–20 per cent of our food, and the time to shift is now: 'We need to be pre-emptive and change the growing culture now and not be trying to do it when we are hungry.'

These impediments to the rise in urban agriculture could also go some way to explaining the class divide in home growing. It takes time, which may be more plentiful for those who are wealthy, supported by a partner's income or retired. For those on lower incomes, especially in more recently developed countries, there may be a recent memory of *having* to toil in the soil to put food on the table: it is seen as drudgery not a lifestyle choice. There is also a question of awareness: the link between food, health and environment is oft repeated only in the kind of middle-class media that is less consumed by the hard-up.

But Dr Tomkins works in some settlements of great hardship and deprivation that completely buck this trend: 'The yield the women are getting off this community garden we set up is 80 tonnes per hectare of green vegetables. The women weigh everything and have become experts in growing tomatoes, lettuce, okra and herbs.'

He is talking about the Domiz refugee camp in the Kurdistan region of northern Iraq, home to about 70,000 people mainly from Syria. He works with the Lemon Tree Trust, a charity that supports gardening and food production in refugee camps. When they were invited to help out in Domiz they were pushing at an open door because so many dwellings already had green space in hidden courtyards or climbing up walls. The Trust supported existing gardeners and encouraged expansion through the funding of a nursery for tree saplings and seeds. Many residents had come from farming backgrounds and had skills to share, while others simply wanted to beautify the house or the camp: 'They plant roses, herbs, broccoli, vines, grass, and even keep water gardens. People grow them for all the different reasons that they grow gardens in Britain.' But he says there can be an extra motivation in a place where they lack much influence over the course of their lives: 'A garden is a place where they have some control. They are "placemakers", where they can create something that is beautiful. (Alongside plants, they are masters with concrete, they sculpt with it.) It's not just about food, it's also recreation, health, trauma recovery. So many things come into conversation when we allow food to happen. They are trying to find a way to self-heal and a reminder of home.'

More recently in Domiz, the Lemon Tree Trust has helped out with the creation of a communal garden with polytunnels that, together with people's individual plots, provide opportunities for incomes and a new refugee economy. It has also tried to engineer ways of using 'grey' water (used in the home but not contaminated with human waste) on the gardens. Normally this is moved out of the camp swiftly but expensively; however, the Lemon Tree Trust is promoting its reuse to irrigate home gardens, market gardens, agroforestry and orchards. An average household creates enough wastewater throughout the year to quench an average garden.

I ask if there is any use of nightsoil (human excreta) for fertiliser as, purely practically speaking, that might seem like an available plant nutrient on the doorstep. But he tells me that is pretty much impossible within Islamic culture, and research backs up his perception. A paper entitled 'Health Aspects of Nightsoil and Sludge Use in Agriculture and Aquaculture' from Cross and Strauss says that: 'Excreta and urine … are regarded as spiritual pollutants by Koranic edict, and Islamic custom demands that Muslims minimise contact with these substances. The use of human excreta in agriculture and aquaculture and re-use of wastewater are not condoned in Islamic society.'

Even without challenging religious taboos, host governments can be hostile to gardening in camps, especially tree-planting, as they see it as evidence of permanence – quite literally 'putting down roots' – when the idea of a refugee camp is for temporary refuge. But Dr Tomkins says the truth is that many of these settlements have now been there for at least a decade with no end in sight, and trees would be an asset even if the camp was abandoned. He thinks everyone involved in the management of displaced people needs to understand the power of plants: 'Why don't we place agriculture and food at the heart of crisis response? Give refugees seedlings and trees. If you visit the camps you will see that people often bring seeds with them from home when they arrive. Many families are in camps for twelve to fifteen years. Gardening provides some self-respect.'

Some cities have tried to make urban agriculture mainstream – such as Havana, in Cuba, where Western sanctions against the communist-leaning Castro regime made food imports scarce and expensive, or Mexico City, which boasts around 250 productive rooftops and peri-urban floating fields dating back to the Aztec period – but I think the best recent example is Rosario, Argentina's third-largest city with one and a half million inhabitants. The country's economic collapse in 2001 left a quarter of Rosario's workforce unemployed and half its population below the poverty line. Food shortages were frequent and prices high. The authorities responded with the Urban Agriculture Programme (Programa de Agricultura Urbana, or PAU). They began with tools and training in agro-ecological practices and expanded to incorporate forty schools and 2400 families. The programme used seven Parques Huerta, or Vegetable Garden Parks, and various smaller neighbourhood plots that were formerly under-utilised or abandoned. This was initially achieved with maps very similar to Dr Tomkins' projections: detailed city plans that identified vacant land unsuitable for other purposes and reimagined those spaces as small farms. They also set up local markets to sell the produce, which they claim is fresher and cheaper than imports and with a far smaller carbon footprint. Today, 300 farmers have temporary ownership of public and private land, and two-thirds of those who grow and sell are women. They harvest 2500 tonnes of fresh produce every year. The growth in lush green space also reduces air pollution and helps to regulate temperature. More recently they have expanded the programme to include a green belt around Rosario of agro-ecological fruit and vegetable production.

However, places actually doing urban agriculture at scale are rare. Internet research reveals mostly academic papers saying how important it *could* be in addressing food injustices. Other top search returns are eye-catching innovations like Zinco, high-tech green roof systems made by a German civil engineering company. There are also revolutionary 'agri-tecture' designs on the drawing board that fully integrate plants and living space, but my personal favourite has to be the proposed metro farms in the South Korean capital of Seoul, where LED-lit vertical hydroponic racks on the actual subway platforms grow salad crops that commuters can eat in the cafe beside the 'farm'. These projects are all plenty of fun but they are not really feeding the world.

Also notably absent from the urban farming scene is much mention of growing the big calorie and protein crops like wheat, beans, maize and rice. Possibly potatoes come closest as they grow well in tubs and their yield of nutrition per square metre is very high.

Despite these current limitations, urban agriculture will still be relevant to our food future. The kind of fruit and vegetable crops you can grow alongside, inside or atop buildings are crucial for a flavoursome and healthy diet. At the moment most of us don't eat enough of those fresh foods and, when we do, we find them easier to buy from the shops whose stock is often grown in distant fields. If or when we *need* to feed from our doorstep, the seedling solutions are waiting to flourish.

As I completed this book, the City of Hull announced it has become the first urban council in the UK to grant residents the 'right to grow' on disused council land. This is political recognition of the argument throughout this chapter: that smart use of the urban landscape includes food production. It helps both people and wildlife and relieves some pressure from the land elsewhere.

2

ENERGY, PART ONE:
SOLAR

A solar farm is the opposite of a regular farm: it appears shiny, silvery, angular and static, not green, soft, matt and subtly shifting. PV panels are modern and human-made; fields are traditional and, at least partially, natural. Solar panels are very good at turning the sun's light into energy; plants are not. It seems rude to say, given that photosynthesis is the foundation chemical reaction of life on Earth, but it's a bit rubbish at creating usable power: it captures around 1 per cent of the sun's energy in plant matter, whereas PV panels can capture 20 per cent and rising with innovation.

I am giving over a chapter to solar, not because it is the biggest land user in the energy sphere – currently less than one thousandth of UK land is under PV panels – but because it is one of the most eye-catching and controversial. Crops grown to make liquid fuel or energy in bio-digesters take up much more land but they don't stick out like a sore thumb or a sea of tin foil. Solar farms represent a fundamental challenge to our concept of desirable landscape and land use that we have developed over centuries. This idea of an alien presence was deployed brilliantly as a protest tool by the Campaign for Nuclear Disarmament in the 1980s, when they were opposing the deployment of American nuclear missiles in East Anglia. They produced a reworking of John Constable's famous Flatford Mill painting where a haywain crosses the mill race, with the dry grass in the wagon replaced by an arsenal of cruise missiles. I may have given anti-solar groups a campaigning image idea.

Botley West is the biggest solar project under development in the UK at the time of writing. It is just to the west of Oxford and, with its three neighbouring sites, totals 1400 hectares or about 2000 football pitches or the size of Heathrow Airport. Whatever metric works for you, it's big: enough to power 330,000 homes, which is roughly the number of dwellings in the county of Oxfordshire. It would add 6 per cent to the total solar-generating capacity of the UK. We need to at least double that total capacity by 2030 according to the Climate Change Committee, the independent body set up by the government to monitor our progress and advise on policy needed to hit our climate goals. The government itself targets a five-fold increase in solar energy by 2035, but that is 1400 hectares of farmland – enough to swallow up seventeen average-size UK farms. It is difficult to calculate the lost food production but, if you assume a fairly typical mix of wheat, barley and oilseed rape in equal proportions and take the average yield for each, the loss would be 3700 tonnes of wheat, 3100 tonnes of barley and 2000 tonnes of rapeseed. The developers argue, correctly, that solar delivers both more money per hectare and more energy per hectare whether measured in calories or kilowatts. But we can't eat cash or electricity.

That is a lot of figures to take in and, while you can't put a number on beauty or heritage – as important as they may be – it is threats to those that really seem to energise the opposition. One of the people campaigning against the site, Frances Stevenson, points at a field and claims: 'It will be an ocean sea of glass and steel. People live here; this isn't a desert in China. Can you imagine going for a walk amongst all that?' When it comes to landscape, people almost always love what they know; change, especially on this scale, is viewed with suspicion. Whether they moved to the area or grew up in the countryside, being in the midst of a solar farm is not people's idea of a rural idyll. But most protest groups acknowledge that climate change *is* a big problem, so what are their solutions?

The first, hinted at above, can be summed up as 'somewhere else is more appropriate': build them where there is more sun and fewer people like the Sahara or the deserts of China or the arid plains of Spain. It is certain that such developments would generate more electricity per hectare and probable that they would anger fewer locals, but it is hard to avoid the conclusion that this is an intellectual veneer over a familiar core: 'not in my back yard'. It puts the responsibility for creating essential green power somewhere else and on someone else.

Their second solution goes to the heart of this book: put them on top of buildings. Cover our homes, offices, barns and warehouses with solar panels. It's an idea I like so much that there are some panels less than a metre from my head as I type. My home has as much PV as permitted, but we'll dwell on the built environment later in this chapter. The scale of the solar roll-out required to significantly slow global warming is so large that it cannot be *either* roofs *or* fields. It will need to be both. The Climate Change Committee and the 'Net Zero Review', a government commissioned study of our climate policies, both reached that conclusion. So how can solar farms cohabit with other critical land uses such as wildlife, food and carbon storage? Let's look at wildlife first.

The Botley West solar farm development claims on its website that it 'presents a rare and fantastic opportunity to bring about significant environmental gains in Oxfordshire… a meaningful net gain in biodiversity across the site area.' Sounds great but, in truth, almost any change in use from intensively cropped and chemically enhanced farmland would be better for nature. It's estimated that halting the use of herbicides, pesticides and nitrate fertiliser would lead to a 10 per cent biodiversity gain in the first year and gradually more in the future. The planning consultants say a 50 per cent gain is well within reach.

And size matters. There is great concern among wildlife experts about fragmentation of habitats. It is difficult for plants or animals to thrive with only small, separated islands of biodiversity either remaining or newly created. 'Corridors' and 'interconnections' are all the rage in nature recovery circles. The scale of Botley West provides the possibility of big swathes of wildlife rich grassland. They may not *look* like a rural idyll as they are striated by banks of solar panels, but I doubt the cornflowers or the butterflies would care. It isn't full rewilding between the panels, as some maintenance is required to prevent shading from foliage or trees taking root, but it's definitely low input from the land manager: an approach that should be species-rich yet cheap to deliver.

Four-legged mowers are also being considered. A local grazer who owns 3000 ewes but no farmland wants some of his flock to go in and dine. The developers like the idea but have been told that mixing with sheep requires the panels to be 40 centimetres higher than planned to allow good clearance underneath. This means raising the top edge from 1.8 metres (a similar height to an average man) to 2.2 metres, and this is not permitted by the planners due to increased visual intrusion, except when the panels are well away from public view.

Within the site there will also be 400 hectares – equivalent to four sizeable British farms –completely without panels. This is made up of 15-metre margins around each field and some land already set aside for nature. Much of this is a requirement from the planners and, while this isn't strictly multifunctional with energy generation, much of it wouldn't be available to other land uses if it wasn't for the solar farm. The developers have engaged Cherwell Collective, the local environment, food and social justice group founded by Dr Emily Connally and featured in chapter 1, to advise on how to use some of this space. So I contacted Dr Connally to find out more about their aims, and she is not short of ambition: 'We are suggesting planting green manures [fast-growing plants to cover the soil that make fertiliser when cut] under and around the panels and then a food forest on the outside. Actually, for the first few years, the best thing for the ground around would be to leave it fallow or plant hemp, which improves the soil condition and adds carbon, and then some clover to get some more nitrogen in.'

When edging a solar park, trees would be planted on the outer rim of the perimeter to avoid undesirable shade. It would take plenty of work to get going but Dr Connally thinks this is possible with some of the 38,000 people in the area who are engaged with the Collective. Once established, she insists, it is relatively low maintenance and would yield nutritious food and a good habitat. I question whether the developers and investors they answer to are genuinely sincere about these collateral benefits alongside the revenue generating panels.

'I'm sure they are sending the sincerest ones to talk to me!' Dr Connally replies. 'But I am not as cynical as some about the involvement of big money. Like it or not, we need very wealthy people to get behind this transition. We must find ways to help them do it and hold them accountable. I hope it goes through. It could be a brilliant thing, a test case to show what is possible: clean energy, high-nutrient foods, greater local self-sufficiency and improved wildlife.'

In late June 2023 I am walking in Devon with difficulty. Each step is harder work than it should be but more rewarding too. The opposing force isn't boot-sucking mud or an implacable headwind but a chest-high meadow of wildflowers. As I push through with ecologist Guy Parker, a bow wave of butterflies advances in front of us, while further ahead we see the movement, but never the hide, of something bigger scuttling through the cover. To our right is a rambling hedge on the way to being a linear wood, and to our left are waves of silver solar panels.

We are at Creacombe Solar Farm, which covers roughly 11 hectares about 11 km (7 miles) east of Plymouth and has a generating capacity of just over 7 megawatts: enough to power around 2100 homes. Clean energy is the priority, but more wildlife comes a close second. This land really is tackling the climate and nature crises. Guy, who founded the consultancy Wychwood Biodiversity, helped design the site. And he loves it: 'You can see fine grasses, bird's-foot trefoil, purple knapweed, ox-eye daisy, red clover, smooth cat's-ear and plenty of meadow brown butterflies alongside the occasional peacock butterfly. You're taking a tired old piece of land and you're turning it into something much more wildlife rich and you've got the solar farm. So it's a kind of win–win that makes me feel happy.'

It all starts with the grasses, he says: 'What we did that was different to normal was sowing the site with a fine grass mix. It allows the wildflowers to come. Then, when the site was completed, all the mess had been taken away and it was just a quiet solar farm, we planted a whole lot of wildflower seed and spread green hay.'

Green hay is the botanical equivalent of reintroducing live animals to a place where they have become extinct, and it's got to be fast to work. First you cut a meadow elsewhere while the hay is green, not dried out, so that you don't damage the 'parent' site. Then the cut hay must then be baled, transported to the new site and opened as fast as possible. 'Speed is of the essence,' Guy tells me, 'so you have to have somebody ready to cut one site and somebody ready and having prepared the soil on the other site. In this case, we had two green hay bales on this strip in front of the solar panels and we rolled them out and we had to use pitchforks. It was very old school to knock it about and spread it across the whole area.' Two summers on from when it was laid, he is happy to see the species-rich meadow is not only well established around the panels but also spreading on to the margins.

This solar farm contains a range of habitats: directly beneath the panels it tends to be damper and darker, favouring thinner grass and a greater likelihood of brambles; between the panels it's more straightforward grassland as access and mowing are critical; like a floral ribbon round the outside is the meadow I've described; beyond the fence, but still within the property, is a different flower mix again, including cranesbill and St John's wort. Less good for panels are a few of those spiky teasels, which grow tall enough to shade them, though birds like yellowhammers love the seeds within.

'As a habitat,' Guy says, 'it's fantastic. It's full of bees and butterflies. It's got lots of wildflowers. The only problem with it is the docks [dock leaves]. And it's quite hard to deal with those because you need somebody to come in and really just chop them down before they seed and take over.'

There is a legal category of 'injurious weeds' like thistles, common ragwort and broadleaf docks that a landowner is obliged to remove if they risk spreading beyond their land. That might sound simple; indeed, Guy says he would be capable of just coming in and doing it in a day, but he wouldn't be allowed to do so. The site has a landowner, an asset owner, an asset manager and then a series of subcontractors for different jobs, so the reality is much more cumbersome and the job could take months: 'To get a message down to the person in the white van who's coming here to do a job may need four or five people talking to each other first. It's classic Chinese whispers. I think it's a major problem with the solar industry.'

As we'll see, this fractured responsibility has negative impacts well beyond meadow care, but for now let's stick to what the good developers can do who incorporate wildlife benefit from the start. The first thing Guy identifies is to treat solar-generation problems as nature opportunities. If you have a big, protected tree that shades an area, or some archaeology that you can't cover with modern technology, then he suggests: 'Let's put some scrub in. Let's build a massive pond. Let's put in a really cool wildflower meadow that wouldn't be suitable elsewhere. That kind of approach is really useful because you're making the most of a site right from the start and you're finding the easy wins.'

The varied, mosaic habitat is loved by conservationists as it helps a broader variety of insects and animals. Wychwood Biodiversity was a partner in a study commissioned by the trade body Solar Energy UK to get good data on nature. Researchers walked set lines – transects, in the jargon – and used random quadrants so they could count their findings without bias, and this standardised approach is now being taken up nationally. As you might expect, the study showed a greater richness of nature than on typical farmland that had been replaced by solar farms, and there was particular species breadth in the deliberately seeded areas. Guy is happy with this but he thinks it could go so much further if the industry really got behind promoting nature, rather than just enjoying the happy accident that almost any development can boast more nature than an intensively farmed field.

'If you could replicate this kind of thing [at Creacombe], then you could go a long way towards reversing some of the habitat disasters we've had,' Guy says. 'Remember that wildflower meadows are some of our most destroyed habitat, with just 3 per cent left of what we once had. The birds [here] are going through the roof and you get a lot more pollinators because of improved botany. If you imagine that achieved on maybe even half of the solar farms across the UK, you've suddenly got a huge area of habitat like a series of stepping stones across the country. You'd have all of these areas where there were lots of pollinators, and some of them may be quite rare, and they would have this ability to spread between the solar farms and into the surrounding landscape.'

One hurdle is cost. Guy reckons on this size of field the initial outlay for promoting biodiversity might be low tens of thousands of pounds a year for a couple of years and then one or two thousand a year for maintenance thereafter. But some sites are ten times as big and the upfront costs may then run into hundreds of thousands. Of course, the income is bigger too, and I don't think this seems like a prohibitive amount of money. But if it isn't a priority and you are competing to maximise returns for investors, then it would be tempting to cut that cost. The way round this would surely be to make higher-level biodiversity management part of the planning permission: you don't get the go-ahead without it.

According to the developer of this site, Harry Lopes from Eden Renewables, that is beginning to happen, as more planners are insisting on a robust biodiversity improvement plan. Harry is developing more multifunctional solar sites, some combining with nature, others with sheep and some with both. He's been in the business for more than ten years and has seen a lot of changes.

'Wild changes in the subsidies and the grid price of electricity make this business a "solarcoaster"', he tells me. 'But it was always obvious to me that combination with nature needed to be done. You are managing an area with very few human visitors, with singular control for twenty-five to forty years and you have to pay some natural maintenance costs anyway. The extra cost of doing it right by wildlife is marginal – it is a total no-brainer. When we started back in 2010, we saw that the [solar] pioneers mowed, sprayed or even removed the topsoil and we thought that was a bloody crazy state of affairs.'

He agrees with Guy that too many different parties involved with a site can hamper good nature plans – as, of course, can owners who just want absolute maximum returns. Harry's company goes to great lengths to make sure the nature-improvement plans are legally locked in for the duration. At the time when I talk to Harry his company is about to install a newer type of panel that delivers more energy and plant growth from less land.

'Single axis tracking' panels follow the sun, facing easterly in the morning, pointing up around midday and tilting to the west in the afternoon. They provide around 30 per cent more electricity for a given area of panel and, because the movement alters the shading on the ground, they also allow more vegetation growth whether for nature or farming. The mechanics needed to rotate the panels make them a bit more expensive, but Harry believes their energy yield and land-use efficiency make them a good partner of arable farming: space them widely enough apart and fat machinery will fit in between.

It is perhaps unsurprising that solar developers dwell on the sunny side of the PV in fields story, but what about the academics who study their impact? Dr Alona Armstrong from Lancaster University runs SPIES, or Solar Park Impacts on Ecosystem Services (I bet they were happy with that acronym), which studies the reality and potential of what solar farms can do for wildlife. She says: 'I am in favour of doing it in a way that delivers as many benefits as possible. It has such potential precisely because it demands space and means we have to think about how that land is best used.'

SPIES identifies five core assets that solar parks offer wildlife:

- they are secure, away from the interference of other land use and with limited human access
- they are long term: most solar land leases are for between twenty-five and forty years
- the land use is paid for and so largely insulated from

commercial pressures that might threaten emerging
habitats
- they are usually on low-grade agricultural land where
  the baseline of nature or carbon prior to installation
  tends to be low, meaning there is much room for
  improvement
- the panels shade the land, making it cooler, damper
  and offering lots of different microclimates. Ecologists often champion the idea of mosaic or patchwork
  habitats, and solar parks can promote that, albeit in a
  rather regimented pattern.

Dr Armstrong's group has mapped all the approximately one thousand solar farms in the UK, not so much
to see where they are as to reveal what is inside the fence:
'Only one third of that area is "over-sailed" by panels.
We made up that word, as "covered" isn't accurate. The
rest is the space between, access tracks and the margin
around the outside but within the park boundary.' The
fact that only a third of the area within solar farms is actually gathering energy from the sun might seem a waste
for an author obsessed with efficient land use. But the
access between is required for maintenance and cleaning;
it can't be a solid sea of PV. The buffer zone around the
outside is needed for machinery like tractors and trailers
to manoeuvre, and the tall hedges round many farms to
shield them from public view also casts a shadow of lower
solar performance so the panels will be set back. In effect,
this is required open space for the efficient operation of a
solar farm. As open space is an operational requirement

of solar, using that space wisely offers an unavoidable multifunctional opportunity.

Measured by roadside protest posters, the most common objection to solar farms is that they are replacing food. If the sun is falling on panels, so the argument goes, it is not falling on plants. But how true is this? If PV panels really were arranged as a solid skin over the land, no light would penetrate and nothing would grow; however, in practice, the demands of angle and access mean there is always space between the lines of shiny silver-black glass.

Cornishman Andrew Brewer believes he can get lamb, more grass, more wildlife, more carbon and more megawatts from his land with the application of solar panels. Wishful thinking? Naive 'cakeism'? Andrew was named 'Grassland farmer of the Year 2022' at the British Farming Awards, so at the very least he deserves a hearing. It's not just his prizes that make him an exceptional farmer, it's also his approach. He believes in taking breaks from the grind for both thinking and not thinking. There's a well-worn gag that if you ask most farmers where they went on holiday they'll reply: 'Rumania'. Say that in a stereotype rural burr and you get 'Remain here'. Many farmers take pride in being lashed to the grindstone 365 days a year. But not Andrew Brewer: 'My wife, Claire, says holidays are cheaper than divorces.' His formula for a successful business shows similar respect for giving yourself time for perspective. He reckons he has three different pay rates: 'You have your day-to-day work milking cows; let's say that's £10 per hour. You have your £100 per hour management time and then you have your £1000 per hour business development time.

That's when you think. So spend more of your time on the £1000 per hour stuff.'

His thought process has delivered: though he comes from a farming family near Newquay, he has quadrupled the size of the holding since he's been in charge. It is now 400 hectares of beef, dairy and a few sheep. His real passion is delivering better grazing with more carbon storage and less chemical input (which we'll look at in chapter 5). He has already got a 500 kW wind turbine (pretty big by farming standards), which barely competes for land at all, and has now got planning permission for fifty times that power from an 80-hectare solar farm. But it will share the field with sheep.

'I believe we can use the same grazing methods and deliver almost as much production as if the panels weren't there. There will be a reduction in the herbage yield but, because the panels provide shelter from the rain, heat and wind, the animal welfare is better so growth rates are better. A hundred per cent of the time the animals will go under the panels to ruminate and sleep.'

Watching his own grass grow has shown that, at some times of the year, it's taller in partial shade as it experiences less evaporation so has more water for growth. As a lifelong stockman, he also knows that sheep get stressed in weather extremes and seek shelter. And there is a growing body of academic work to back up his lived experience. A 2021 study from Oregon State University in the USA compared lamb growth and pasture production in similar ranges, one with panels and one without. Though there was less forage in the solar pastures, the quality of that forage was higher, so the study found that the lambs gained the same amount of weight in the two plots. The shade provided by the panels provided a refuge for the lambs on very hot days, conserving their energy and improving welfare. The study also noticed a longer flowering period for plants beneath the panels. Overall, the return from grazing was $1,046 per hectare per year in open pastures and $1,029 per hectare per year in pastures with solar panels. It's worth remembering that the sward in the solar farms had not been planted with more shade-loving varieties, and – most importantly for a land manager – that the figure above doesn't include the electricity-generation payments or the saved costs of regularly mowing or spraying herbicide in a solar farm.

Cornwall is popular with solar farm developers because, together with a strip along the south coast, it's the sunniest part of the UK. Couple this with its popularity as a holiday destination and undeniable beauty and you have a recipe for friction against new solar farms. Local objectors reach for the food argument, saying, 'Don't let an energy crisis today turn into a food crisis tomorrow.' In one sense, Andrew Brewer is sensitive to this, arguing passionately against large-scale rewilding and the consequent loss of food production.

'I think rewilding is a terrible waste,' he says. 'We are in a human-made, farmed environment. I think it is a very big waste of our potential to feed our country first and the rest of the world with a growing population. If I left this farm, in five years we would be covered in willow trees with less wildlife. From now until 2060, we need to produce more food than we have produced in the history of the world to date.* So the need to feed the human population is going to be the driver of all political direction going forward. So why are we giving up productive land on our doorstep? It's almost a national security issue, giving up these areas for rewilding.'

But rewilding or even abandoning highly productive farming is not what is happening here. His solar farm will be a *solar … farm* doing both. 'I will still be farming that land, people will still be nourished by that land and more wildlife will be thriving on that land.' If we had a metric for land productivity that measured all the benefits to society, his patch would be topping the scale. That is what an unusually original and ingenious farmer can do with an otherwise typical West Country farm.

Data from January 2023 reveals there are roughly 1.2 million homes with solar panels across the UK. That is a little over 4 per cent of the housing stock. In the US the

---

* I hear this claim a lot and decided to check it out. World population today is just over eight billion and is predicted to rise close to eleven billion by 2063. Let's make a generous assumption that the average population in the next forty years is ten billion. Let's also assume that someone in the wealthier world consumes an average of 3500 Kcalories per day, so around 1.3 million calories a year. 40 × 1.3 million × 10 billion = 511,000,000,000,000,000 Kcalories. The number of people who lived between 8000 BCE (a reasonable date for when most people ate food from farming) and today is about 110 billion. Let's assume they lived for an average of thirty-five years and ate less – say, 2500 Kcalories per day (although there is some suggestion that peasant manual labourers in the past might have eaten more, possibly 4000 Kcalories per day), or 912,500 Kcalories per year. 35 × 912,000 × 10 billion = 3,513,122,500,000,000,000 Kcalories. This number is nearly seven times larger than the first, so the claim is wrong: we have eaten close to seven times as much in the course of human history as we will in the next forty years. But we still need to grow a staggering amount more each year than the historical average to feed our enormous population, especially if we maintain or increase meat intake.

proportion is similar. The European Union is seeing rapid solar roll-out, with more than half of it on domestic rooftops rather than utility scale. Deployment has been boosted recently by climate concerns, the high cost of energy and fears over security of supply, the latter two both worsened by the war in Ukraine (solar installation rates in Portugal, for instance, more than doubled in a single year between 2021 and 2022). But in order to understand the rollercoaster roll-out, we need to look further back.

In the early 2010s Britain – like much of the rest of western Europe – wanted to encourage solar panel uptake on homes, so the government offered a payment on every watt generated by new installations on grid-connected homes, up to a maximum capacity of 4 kW. Confusingly, this was named a Feed in Tariff (FiT) but it had nothing to do with how much you supplied – fed in – to the grid as, even if you used your entire production yourself, you would still get a payment that started at 41.4 pence per kWh, which was linked to rise with inflation. You did then get a small amount on top of this – 3.1 pence – for every kWh you sent to the grid, but this was estimated, not measured, and set at half your yield. It wasn't actually a payment from the government, from taxation, because it came out of everyone's energy bill as a 'green levy'. The final economic advantage for homeowners was lower electricity bills, as customers with solar tend to draw less from the grid.

I'm lucky enough have a south-facing roof on a garage/home-office building, and I had panels installed under this scheme in 2011. Since then they have worked without fault and generated 50,000 kWh. Let's get a better handle on that number. It could power a 100-watt bulb (like the old-fashioned tungsten filament ones) for fifty-seven years, a typical 800-watt microwave oven for over seven years, or keep a 3-kW kettle bubbling for a little shy of two years. That amount of electricity would shift an average electric car 281,635 km (175,000 miles) – seven times round the Earth. And, if you prefer a generation comparison, it is the same as 50 MWh (megawatt-hours), which the most powerful single wind turbine in operation today could deliver in about three hours on a windy day.

So much for the power; what about the money? Over those years I have been paid more than £20,000 and this will continue at a rate of more than £2000 per year until 2041. There has been much criticism of this solar incentive scheme for being too generous and socially iniquitous. The latter is because all bill payers, including those on low incomes, are paying panel owners who tend to be wealthy, and will continue to do so for thirty years after installation. One estimate I have seen from Green Business Watch calculates that over the thirty-year period my total income and saving will be around £66,000.

The point of these initial incentives was to create a market in domestic solar installation so that capacity of the nation's roofs to generate renewable energy wasn't wasted. Early adoption of any new system is often perilous so the pill had to be considerably sweetened. Nevertheless, the fact that I and many others with the right building geography and bank balance saw this as a sweet deal led to very rapid uptake, and the incentive level was halved after two years. But during that time, big institutions saw a chance to profit from small homes. At one end, commercial companies offered to pay you rent for the space on your roof where they then fitted panels at their own expense. You got a small annual payback and the electricity for free, they got the fat Feed in Tariff. So now the shivering pensioner in her hard-to-heat home was paying financial institutions too. However, the other group that spied the opportunity were housing associations and local authorities, who covered much of their stock with panels and offered a similar deal to the families within. So it did help some providers of social housing.

A couple of figures here might be useful. The installation of my panels in 2011 cost £15,500. In 2019, I put in the same number on a different stretch of roof when all incentives had pretty much gone. That cost me £6,000 and the roof was less accessible so mounting them was dearer. Some of that difference is explained by panels themselves becoming cheaper, but a large chunk was going to the early installation companies who were able to charge more because the subsidy was generous. The Feed in Tariff continued to decline in stages until April 2019, when it ended altogether. Domestic solar can now compete without subsidy.

On reflection, I think the initial payment was too generous and should be paid for out of taxation rather than bills, but it did spark a domestic solar revolution.

Fast-forward a decade and we are in another solar deployment boom. According to the Microgeneration Certification Scheme (MCS), which certifies small-scale energy technology, there were 21,000 installations in the UK in the month of January 2023 alone – double the same month in the previous year. Gareth Williams from the renewable power developers Caplor Energy says the reason is clear: 'The economic case for rooftop solar is mind-numbingly obvious. Even in the subsidy-free world of today, payback on installing panels is four to five years. Less – two to three years – if you use a lot of electricity yourself. Our company has grown two and a half times the size in the last year and we could have probably quadrupled if we could find the staff.' He is mindful not to sound too triumphant, given that war is an accelerator of this boom, but money talks. MCS forecasts for 2023 were that more than a quarter of a million UK homes would have solar panels installed in a single year.

The logic of combining living room with energy generation is proving inescapable. But what about where we store and make stuff, our industrial buildings? Of everywhere under the sun, one place gathers almost universal approval for the location of solar panels. Thankfully, this is a growing space, frequently visible beside motorways and especially common near junctions. It is the unavoidable, giant box of our delivery-addicted lives: the distribution warehouse.

'This is a 520,000 square foot warehouse,' Lindsay Mackie says as she welcomes me to the biggest indoor space I have ever been in: Wincanton Distribution Centre in the Northamptonshire town of Corby. It is 12 metres high, mostly occupied by giant metal shelving, and it could accommodate nearly seven football pitches. The combination of its enormity and the apparently endless racks disorientates me. As my eyes search for the end point and my brain struggles to compute the novel visual data, I feel like a movie robot discovering something new. It's like horizontal vertigo.

This distribution centre specialises in hardware, so the mechanical finders and the staff are busy retrieving saw blades, drill bits and sandpaper for home delivery or collection from a large range of DIY stores. They have hundreds of huge boxes, each the size of half a car, filling the giant racks. Some contain solar panels, while there are other boxes with batteries and yet more with the frames to mount them on roofs. But there is none of this renewable energy tech on the roof of this building, an area about 1500 times the size of my own domestic solar array. Why?

'Personally, I think it's an absolute travesty that it is not covered in solar panels,' says Clare Bottle, CEO of the UK Warehousing Association, who was on my tour of the warehouse floor before we found a spare conference room for a chat. 'But we probably have dozens of people who think that it's a travesty. That's not going to solve it. What I am trying to think about is how as a trade body we can make it better.'

Clare opens her laptop to show me the Google Earth view of Corby. A town in central England surrounded by trading estates, it is blessed (or blighted) with plenty of warehouses.

'There are quite a lot of industrial buildings and, if you look, you can't see any solar panels; they are just not there. If you zoom in, you can see a lovely bit of real estate just crying out for solar panels, but they are not there.'

We then scroll 128 km (80 miles) north, and Clare continues: 'This is Next [the clothing retail chain] in

Doncaster and they have covered much of their roof in solar panels. It's not a blanket across the whole lot but most of it. If you then zoom out, you can see their neighbours haven't done it.'

Our eye in the sky then whisks us down to Magna Park, close to the junction of the M1 and M6 motorways. It claims to be 'Europe's largest dedicated logistics and transport park'. Their website boasts they already have over 1 million square metres of 'sustainable floor space across 41 buildings' and, having acquired more land to the north and south, will soon be able to offer 1.5 million square metres. That is more than thirty times the size of the warehouse I visited in Corby.

'There are no solar panels to be seen,' Clare points out. 'There is one with a dark roof here, which may have a solar film on. This is the biggest number of warehouses in one place and just to the south of it is DIRFT – the Daventry International Rail Freight Terminal – more warehouses and none of them have got solar panels on either.'

One third of all commercial roof space in Britain is on warehouses and that proportion is rising. The UK Warehousing Association (UKWA) has calculated there is room for 15 gigawatts of PV, which would double the country's current capacity. Just the largest 20 per cent of warehouses cover 75 million square metres, equivalent to the footprint of half a million average houses. Given that electricity prices doubled in the years 2022–23, the potential to reduce energy costs is huge. Overall, the UKWA reckons rooftop solar PV has the potential to save the industry £3 billion per year from a combination of reducing their own electricity bill and selling power to the grid. The former is always more lucrative as, on average, you pay four times as much to buy power than you get for selling it. Deployment at such scale would also make the industry carbon negative and burnish their green credentials.

Put all that together and it seems like the lowest hanging of fruit or, perhaps more appropriately, the easiest 'off the shelf' solution. So why is only a dismal 5 per cent of warehousing wearing a shiny solar hat? Clare, who has made it her mission to make warehouses multifunctional by both storing stuff and powering stuff, believes we can do better: 'For every house in the UK there is 6 square metres of storage space. That's growing and we need to be using it. I want to be the catalyst that triggers rapid solar roll-out.'

There is 50 per cent more warehousing in the UK in 2023 than in 2015, which is due to a range of factors such as the growth of e-commerce and the fact that preparation of goods for home delivery takes roughly three times as much space as organising the same goods for shops. Also, according to Clare, Brexit means it is harder to move stock in and out of the UK so more goods are held in warehouses to protect against those delays.

For all the motives, though, there are also considerable obstacles, which can be split up as grid connection, ownership structures and energy consumption. Let's start with grid connection. For anything but the smallest patch of panels you need permission from the local Distributed Network Operator (DNO), who needs to be sure that the nearby cables are able to take the load and shift the wattage to where it is needed. Given that the British grid was developed when electricity was generated from a handful of massive power stations and then consumed by millions of, mainly small, customers, it is perhaps not surprising that it is struggling to adapt to a new world of multiple generators of varying sizes and wattage flowing both ways. Some projects may get lucky and their new solar-generation requests may receive the DNO's approval with no constraints and no charges, but often, if the DNO's response is not an outright 'no', it varies:

- 'Not yet': the grid capacity is being expanded but too slowly for many potential renewable energy projects, which are told they may not get hooked up for at least a decade. According to Clare, this was the case with a new development at Magna Park that was told there was a queue of a hundred projects to connect to the grid in the area and they were at the back of it, with no grid reinforcement work planned until 2028.
- 'Not that big': constraints on the maximum size and output of the array can make deployment uneconomic and also simply mean less green energy on the grid.
- 'It'll cost you': the DNO will insist that the solar project owner pays for the grid infrastructure upgrade.

Clare doesn't mince her words about DNOs: 'Too often they act like shoddy middlemen. We are often asked for at least £1 million for a connection. They have no incentive to make the investment [in renewable energy] that the country needs. All they are concerned about is the price, and OFGEM [the statutory energy regulator] is not doing enough.'

The problem with the next obstacle – ownership structures – is illustrated by Clare's response when I ask her how it can make financial sense for me to put solar panels on my tiny roof but they are so rare on the acreage of industrial warehousing. 'The clue is in the words "me and my",' she says. 'It is unambiguously your place; you control the thought process, the investment decision. You decide how to use the resulting energy and you decide what to do with any payback.'

In contrast, the warehousing business is increasingly segmented. One site may involve a landowner, a developer, a building operator, and a tenant – all potentially with different investors with different timescales of engagement. The building may last either side of fifty years, the panels around twenty-five years, the tenancy ten to fifteen years and the payback time on the panels four to six years. Who pays, who agrees, who insures and who benefits? And who cares enough to think about installing solar panels and not just concentrate on the core business of making a profit and storing some goods? This is where Clare, as boss of a trade association, thinks she can really make a difference by bringing these people together and getting them pulling together.

'It's really appalling that new developments are going up without solar,' she says, 'but I think attitudes are changing. I talk about solar wherever I go and it "lands well". People are much more interested.'

She puts this down to genuinely increasing concern about climate change and a better business case driven by higher energy prices, falling panel costs and trends in energy usage in warehousing. On average, compared to their size, warehouses have not been big energy users themselves, reducing the financial incentive for on-site generation. But this is changing. Home food delivery demands more temperature-controlled, usually chilled, warehouses that use much more electricity than regular big sheds so they are more frequently adorned with solar. As the relentless switch to automation drives warehousing to run on more robots and fewer Roberts (or Robertas), more power is required from the socket and less from the staff canteen. That means more watts, fewer calories. Cheaper big batteries are also enabling shifting energy from day to night. But the change that Clare is really excited about is less about the storage and more about the transport: 'In the future, as more of the transport fleet becomes electrified, the warehouse could provide the wattage as well as the goods.'

The UKWA's own study suggests that decarbonising one third of the country's heavy goods vehicles (HGV) would need around 15 GWh (gigawatt-hours) – similar to the total amount of solar energy generated in the UK last year – and could be generated from warehouse roofs. When you see all those trucks 'plugged in' to endless loading bays alongside those massive dull grey rectangles, this seems like a really snug fit.

Making useable energy from the light that falls on a warehouse is one of the most obvious and productive multi-functional land uses. The economic and environmental case is sound and becoming stronger. The obstacles, while real, could be overcome by regulation, for instance planning permission contingent on solar and energy being much more prominent in the local plans, the businesses involved being more driven by the long-term opportunity rather than short-term return, and the grid being funded and forced to offer connections. Clare sees a vital role for central government but she's not asking for funds – at least not directly.

'Money isn't really the problem here. There is plenty of private investment money available. One of our associate members in the UKWA is Zestec, a branch of Octopus energy [one of the UK energy suppliers] and they have offered a deal where they fund all the capital for putting the solar panels on your roof. They can also fund some roof re-enforcement work if necessary. They then sell the energy to the occupant. The guys at Zestec are telling me we have shed loads of cash to throw at this. So I don't think money is the problem.'

If you make a big green investment, such as in solar panels, you are improving the value of your property, and in England you get an exemption so you don't have to pay more business rates straight away, but in other UK devolved nations you don't get that. Clare would like to see that tax break everywhere.

I started this warehouse section by saying that the idea of smothering them in panels is popular and, having heard the arguments, I think 'the people' have got this one right. There is a huge, affordable energy-generating opportunity, with no space penalty, on industrial land occupied by businesses with commercial and engineering know-how. I can't think of more fertile ground for a solar farm. Frequent lack of grid connection is the one strong (but not good) excuse, but all the others seem a bit petty given the size of the prize.

There is surely a role for government to bring the players together, find a way through and call out those who are dragging their heels. A popular strategy that doesn't cost money and cuts climate change without using up land: it's fruit hanging so low a toddler could grab it.

Just before we leave the industrial estate, Clare intrigues me with another revelation, which isn't relevant to energy but is highly relevant to saving space: warehouses are getting higher so they can hold more stuff. In 2015 the average warehouse was 11 metres high, whereas new ones average 15 metres. Some stretch to 20 metres but taller buildings enlarge their visual footprint, angering both locals and planners. One way round this is to dig down, and some warehouse floors can be 3 metres below the surrounding land, enabling their holding volume to increase while they look no taller. Another, probably cheaper answer is paint. Clare points out how some units have tried to become less intrusive with graded wall colours chosen to blend in with the sky: 'I think they look great, but I would say that. I think they are really beautiful and can enhance the landscape.' She laughs when I suggest that such beauty might be in the eye of the beholder; she is clearly a woman who loves her job.

All this discussion of warehouses prompts a further thought: another type of building with a lot of roof space and a sizeable energy demand is schools. 'Solar for Schools' is an organisation that seeks to use the top side of the school to deliver both electricity and education. They bring together charitable funding grants and some investment to try to make the installation as cheap as possible for the school but also offer curriculum-based teaching modules for physics and maths. So far, they have installed systems on 168 schools in the UK, two in India and 87 in Germany, delivering a total of 37,500 MWh. In today's consumer electricity prices of 30 pence per kWh that is £11,236,500. They are not the only player; other companies like JoJu Solar and Solar Sense help to bring the total number of UK schools getting power from the sun to well over 1000. But with over 32,000 schools in total in the UK, that still leaves plenty to go. Combining education and energy generation on the same patch of land is surely another no-brainer.

On a summer's day in 2014 Elon Musk was in the UK, promoting his latest Tesla model. The photoshoot was at a parking bay with a charger on the wall and solar panels on the roof. Ole Gregerson, chief executive of the company that built this bay, Bluetop Solar Parking, takes up the story: 'It was greyish weather, the panels were doing very little, and then ten minutes before the photocall the sun came out. Good for him and good for us. But it was a temporary structure only standing for a few days.'

The PR stunt may have been ephemeral, but the idea of solar panel-covered parking lots has got real legs. Ole believes you can cover any outdoor parking with PV and he now has permanent sites in Dundee, Falkirk football stadium, a car charging bay in Essex and in a Leeds park-and-ride. At the time of writing, he is waiting for the go-ahead to solarise 1400 carpark spaces at a business centre on the south coast of England. Across mainland Europe he has solar carports in five more countries and one on top of a multi-storey carpark in Dubai. It is more expensive than regular ground mounting as they are higher and need more of a substantial structure, but they will pay back over time (how fast depends on the price of electricity). But financial cost isn't the only metric, says Gregerson: 'We should make use of all the opportunities for solar as using space efficiently really matters to me. Our vision is to have a solution for any carpark anywhere.'

Many of these carport solar developments are co-located with EV chargers. It feels good to drivers, as it did to Elon Musk, to be fuelling direct from the sun. But is it pointless or purposeful? In a narrow sense it could appear a bit bogus. If you were charging your car on a clear sunlit day at a typical commercial charger speed of 50 kW you would need a solar array around fifteen times bigger than the usual domestic set-up, so much bigger than a parking space to deliver that wattage. But time delivers a different perspective and those panels are continuously hoovering up any solar radiation. Remember that statistic I gave earlier that the 4 kW of panels on my roof, having generated enough energy so far to drive a car for 281,635 km (175,000 miles)? That means that every year they generate enough electricity to power two electric cars doing average annual mileage. Let's see how much you could get from covering a normal 12 m² parking space with PV panels. Typical panels yield 0.265 watts per square metre so if you covered it entirely you would get 3.18 kW. As my 4 kW array gives enough for nearly two cars, 3.18 is easily enough for one. It's true that you wouldn't cover every square centimetre with PV cells, but also true that I haven't included the space for the lanes between bays.

Including this yields another 7.2 m² per car. In summary, the solar panels on top of a parking space generate enough electricity to power a car.

This has got to be a great use of space. I don't really care if they have charging points attached; it is just a good way to generate clean energy to stick in your car or in the grid. And this calculation is for the UK; think how much more you might get in sunnier countries.

Let's go crazy and multiply this up by the total number of parking spaces in the UK. The motoring organisation RAC Foundation states on its website: 'There are between 17,000 and 20,000 non-residential car parks in Great Britain, including those run by councils, commercial parking companies, shops, hospitals, businesses, railway stations and airports, providing between 3 and 4 million spaces. The majority (92%) of these car parks are at ground level. The rest are multistorey.'

This means the space above our parking lots could fuel at least three million cars with clean green energy. That is 10 per cent of the UK's cars. And deployment is increasing. At the time of writing, in mid-2023, Northumberland Council unveiled a huge solar array over their staff carpark at their HQ in Morpeth with 800 kW of generation, 400 kWh of battery storage and 120 EV charge points. These are accessible for both staff personal cars and council vehicles. They hope to use the experience to develop similar solutions for public carparks. France has gone much further with a new law, passed in 2023, saying all carparks with more than eighty spaces must be at least half covered in solar panels in the next three to five years.

I got rather excited about the potential of using the sun falling on carparks and spoke to another developer, John Hewitt of the architects Hewitt Studios. You might argue his approach is more about quality than quantity. He believes passionately in the power of fine design: 'Photovoltaic panels and electric vehicles are great but they are all about tech and that in itself doesn't motivate sustainable change. Since 2014 we have focused on energy: creating environments to hasten the transition to solar power. We want to inspire through experience and architectural beauty to deliver engagement.'

His practice delivers what he preaches. Both the roof and the support structures of his office building are works of art. Their solar roofs are made of glass with embedded PV cells of about 10 by 10 centimetres, and the space between each cell is just transparent glass. This means his panels lose about 30 per cent of the potential solar power from the same area, but that lost light is found below. It gives an open airy quality to the space beneath, which John describes much more eloquently: '*Komorebi*: it is a Japanese word that has no direct translation into English but describes the quality of light that falls through a tree canopy. We are not geared up to maximise production but create an environment of extraordinary character.'

To that end, the building's uprights and beams are made from wood, more accurately laminated veneered lumber: lays of spruce or larch grown at height or far north for closer grain and greater strength. In the solar carport a central trunk then splays out above to support the cells that absorb the sun's energy. It looks like a squat, flat-topped tree and does a similar job: recruiting sunlight for energy. This wooden construction is much more climate-friendly than its steel-framed counterpart but it is more expensive. When I put it to John that higher price and lower yield might not be best suited to a climate crisis, he responds crisply: 'Ugly is not fine in urban environments.' That's me told. Their stunning K:Port EV charging and mobility hubs can be seen at Portishead Marina in North Somerset and beside London's Woolwich ferry.

Once you start to do the research, it's easy to get carried away with the variety and possibilities of solar power generation. It encourages ingenuity. For instance, researchers at Leipzig University in Germany claim great results from vertical panels erected like parallel fences across fields with cells on both sides. As the sun travels from east to west they perform best in the morning and afternoon when other panels are struggling. Also, and importantly for this book, their slim upright stance means they require very little land so it's easy to farm between them.

Another example comes from a Ukrainian working in Germany, Karolina Attspodina, who has founded the company We Do Solar, marketing very lightweight panels that you can fix to your balcony with little more than cable ties and then plug them into your house with a micro-inverter (panels put out DC, whereas homes run on AC). It's a great option for apartment owners who don't have a roof and also renters who might want to 'go solar' and can now take their panels with them when they move. Also in the home, smart tech allows your gadgets to prioritise using the sun's power when it is available (at the time of writing, around £1000 buys you 600 watts of panel capacity and the associated kit).

Elsewhere, floating solar is being deployed on hydroelectric reservoirs, water treatment ponds and even the calm sunny waters of the tropics, and in Switzerland a company called Sun-Ways is rolling out solar panels between railways tracks – a very perilous place for any other activity, but the panels are unperturbed by the occasional train zooming over them.

Solar is an extraordinary global success story. Its growth rates are stunning: in 2023 alone China added more solar energy than the entire installed capacity of the USA. International Energy Agency (the autonomous intergovernmental organisation that provides policy recommendations, analysis and data on the entire global energy sector) forecasts, which have tended to be conservative on renewables, say solar will be generating more electricity than coal-fired power stations by 2027. Although it relies on 'area under the sun', in practice its competition with food for land is extremely limited and its potential for cohabiting with wildlife is huge. Done right, especially combined with buildings, it is very land smart.

3

ENERGY, PART TWO: BEYOND SOLAR

So far I have focused on solar energy but we have myriad power sources, all of which gobble up land. Given that one of the main concerns of this book is efficient use of space, the key metric I will focus on is how much land they take for every unit of power they produce: square metres per kWh per year. The super space-saver is nuclear power and the super-spreader with a large area per kW is, perhaps surprisingly, hydropower. However, bioenergy – plants grown to burn for heat or converted into liquid fuel for transport – has by far the biggest overall footprint, although there is plenty of range within each technology so is it fair to single out energy crops for criticism when solar is getting a fair wind? Let's look at the figures for how much power you can get from a hectare.

The chart below (from Forest Research, part of the government agency responsible for trees) gives various yields. This includes both crops for burning directly (like willow or miscanthus, a kind of giant, perennial grass) and those that are turned into liquid fuels such as bioethanol and diesel (like wheat or rapeseed). Generally, using the whole plant for burning – often referred to as biomass – has better power yield per field than that converted into liquid fuel. That is not entirely surprising as the former uses more of the entire plant. There have been experiments and indeed whole bio-refineries dedicated to turning woody biomass (the ligno-cellulosic ingredient) into liquid fuel but, so far, they have struggled to deliver at an economic scale.

Land use of energy sources per megawatt-hour of electricity*

Fuel	m² per MWh
Biodiesel (from rapeseed oil)	884
Wheat straw	789
Bioethanol (from wheat)	588
Bioethanol (from sugar beet)	303
Wood (SRC Willow)	217
Miscanthus	158
Hydropower (small to medium plants)	33
Concentrating solar (tower)	22
Solar photovoltaic, silicon (installed on ground)	19
Coal power	15
Hydropower (large plants)	14
Solar photovoltaic, silicon (installed on roofs)	3
Gas plant	1
Nuclear power	0.3

* The land use figures for hydropower, solar, coal, gas and nuclear are based on life-cycle assessment and also account for land used for the mining of material used in its construction, fuel inputs, decommissioning and the handling of waste. Whereas for the crop based fuels it just includes the field area where they are grown.

The first thing to point out is that the calculations are based on the full life-cycle assessment of the energy source, so they do not just take into account the area taken up by the generation system itself – the power station or hydroelectric reservoir – but also the minerals used in building it and the extraction of the fuel needed to keep it running.

Nuclear is the most efficient source of energy when it comes to land use because reactors themselves are compact and, more importantly, the uranium mines are small compared to the embedded energy retrieved in each tonne of ore. Both of these points may seem absurd when you think of the massive nuclear projects or vast holes in the ground dug for uranium, but compare them to the number and scale of coal mines. Some may argue that the potential footprint of nuclear is large as an accident spreads its effect over continents (following the Chernobyl disaster in 1986, the soil of 300 farms in North Wales over 3000 km, or 2000 miles, away was still considered too contaminated to allow human consumption of local lamb until 2012), but if we use this pollution metric we would have to say that fossil fuels have an impact across the whole world thanks to global warming. I consider the chances of halting climate change to be much higher with nuclear power in the mix, so I'm relieved that it appears to be thrifty on acreage.

Next most space-efficient is gas, as its extraction requires a small wellhead and not a sprawling pit. I was really surprised when covering a story about fracking for shale gas in America that, after the initial drilling (which may take months), all that's left during years of extraction is a large gas valve – a little over human size – in an open area not much bigger than a few tennis courts. Gas needs refineries too but, overall, its footprint is quite condensed.

Let's consider some of the renewable energy sources. As we have seen, solar on rooftops does really well as the deployment area is counted as zero. As it is multifunctional with other activities like homes or businesses, its land take counts as nothing. When mounted on the ground, not the roof, solar PVs yield for a given area has a wide range as you get more power per square metre in the Sahara than the Shetlands. It seems surprising that hydropower should be relatively space-hungry, but think of all the arguments about flooding valleys behind hydro dams, with towns and villages gone, people displaced and habitats submerged. I would argue that many if not most of these reservoirs are multifunctional as many have provided space for water sports and habitats for wildlife. Once again, there is a range in the figures as it depends on the depth of the reservoir: a deeper one would hold more water for each square metre and therefore deliver more power for each square metre.

But there is something bobbing along that can multiply the space efficiency of both hydro and solar power: float the PV panels on the reservoir behind the dam. Floating or rafted solar is the gorgeously simple practice of mounting panels on rafts, bolting a few together and tethering them on a lake. The pure construction cost tends to be higher than ground mounting but it readily competes where land is expensive, highly productive or simply deemed inappropriate for solar farms. And we have a lot of available water surface. A recent study, published in the journal *Nature*, suggested that covering around one third of the surface of the world's 115,000 reservoirs with panels would deliver three times the total energy production of the European Union.

The biggest floating solar farm in Europe is in the Portuguese region of Alqueva, east of Lisbon and close to the Spanish border. There is a huge hydroelectric dam there and, on the lake behind, four hectares of floating solar generating enough to power 92,000 homes while occupying well less than one thousandth of the lake area. Total coverage could have damaging effects on the reservoir's ecology but at moderate levels it seems to have some benefits, including providing habitat variety and anchor points for aquatic plants. Meanwhile, on the electricity side there are some great win–wins. Shading the water surface cuts surface evaporation losses, so there is more water to power the turbines in the dam. Solar panels are more productive at lower temperatures, and proximity to water has a mild cooling effect. The combination of power from rainfall and sunshine offers a great seasonal balance too: hydroelectric power is often limited in the summer when it's dry and water levels are low, but this is when the sun shines brightly and PV generation would be high. The final advantage is a little prosaic but nonetheless crucial: the wiring is in place. Hydroelectric plants already have the hardware to handle and shift the wattage. It will likely need some modification but that is much cheaper and less challenging in a beauty spot than building new power infrastructure.

A bright spot for this multifunctional solution could be Africa. The continent currently gets 17 per cent of its electricity from hydro, and in some countries like Ethiopia,

Uganda and the Democratic Republic of Congo the figure is in excess of 80 per cent. As Africa develops it will need more electricity but rainfall tends to be seasonal and climate change makes it less certain. In a famously sunny continent, this makes floating solar a particularly smart partner to hydro. At the time of writing, the Ugandan electricity company has awarded a contract to a Swedish company for a pilot floating solar plant.

Across the world, installed floating solar capacity has grown twentyfold in the last ten years. Far from all of this is combined with hydroelectricity. Farmers often put smaller arrays in irrigation pools. I visited one on a fruit farm just to the west of London, where land is highly prized, and the grower was very happy that he could combine a water source with a power source. In the Dutch city of Rotterdam there is a floating array in one of the docks, supplying electricity to the floating farm next door. This farm is a high-tech dairy unit, with a robotic milker, for around fifty cows. It's basically a three-storey modern barn with cows at the top, milk processing in the middle and cheese making on or below the waterline (the even temperature provided by the surrounding water is an asset). Both the farm and the panels each cover an area a little larger than a tennis court. They make and sell cheese, milk and yoghurt with a big emphasis on local and space-efficient food production. They use the electricity from the panels to power the dairy and process the manure into a soil improver. But they do have an area of land dockside that is at least three times larger than what's on water and much of that is covered in grass, which the owners tell me is accessible by the cows as they have their own bovine gangplank that they can cross any time of day. When I suggest this rather undermines their space-saving narrative, they shrug and say: 'The cows rarely cross to the pasture, but customers like to see it. The grass is for the happiness of the people not the cattle.'

To the west of London, near Heathrow airport, solar panels grace the surface of water treatment ponds near sewage works. In East Asia land is extremely precious especially near urban areas, which have a high electricity demand, but these cities are often coastal, and disused docks or shallow lagoons are prime spots for a solar raft. In Taiwan a huge one, covering 88 hectares, sits outside the regional city of Changhua, and what will be the world's biggest is being developed on tidal flats at Saemangeum in South Korea. These are both on salt water, adding an extra challenge for electrical engineers but also opening up the prospect of solar farms moving offshore. It is an intriguing venture, which might reduce pressure on land in the long run, but such marine developments feel a bit too sci-fi for now and outside the scope of this book.

Wind

How do you measure the land use of wind power? Let's start with the tiniest. If you build them offshore the turbine land take is zero except for a small area for transformer infrastructure on the coast. The actual footprint of onshore turbines is very small too. According to Our World in Data, its land efficiency is close to that of nuclear power:

0.4 m²/MWh for wind and 0.3 m²/MWh for nuclear. But this is measuring only the excavated area for the concrete turbine footings themselves. If you count the whole project site, the land take is much greater and highly variable according to the size of the turbines and the windiness of the location. Also, this measure ignores the fact that the space between is almost always used for another purpose like farming, forestry or recreation.

There is another measure of turbine impact on land to be considered: visual impact. You can see them for miles around as they are giants that dwell in exposed locations. Some communities and many politicians consider their addition to the landscape to be so harmful that they are banished offshore at much greater cost (though generally greater efficiency). The National Parks of the UK are virtual no-go areas for big wind turbines and, seeing as park status is predominantly a designation for landscape beauty, it implies they would damage that visual appeal. In Scotland, an additional and somewhat vague description of 'wild land' is also hostile territory for wind-farm deployment. The lack of visible modern infrastructure and abundance of solitude is its USP. Turbines are evidence of human impact and so I can understand how some people could resent that.

Yet I can't help feeling there is a large helping of neophobia here too – fear of novelty. People tend to love the landscape they are familiar with. I encountered a great example of this in Lancashire a few years back while filming a story about wind-farm development for *Countryfile*. We were talking to a farming family about plans for a relatively modest-sized turbine on their land. There was much local opposition around its impact on the view and the fear that it might 'open the floodgates'. Yet also in the area were a number of small mill buildings with tall chimneys dating from the nineteenth century. They were collapsing and a campaign was under way for their preservation. The industrial scars from history had morphed into the landscape treasures of the present, while today's creators of local wealth and clean energy were despised.

Sarah Merrick is the chief executive and founder of Ripple Energy, a community renewable energy developer with turbines in Wales and Scotland. When we speak, her company is just attaching the blades to a wind farm in South Ayrshire. 'Overlooking one of Donald Trump's Scottish golf courses,' she tells me. During construction wind farms have a greater land use, but Sarah says once they are in place the land between them is open for any other business: 'Sheep grazing is the most common as they are often sited in uplands, but there are plenty on arable farms where the cropping extends to just a few metres from the turbine base. In Finland and Norway, they often grow trees underneath and the towers tend to be a little higher as the effective ground level is raised. Overall wind energy uses a tiny amount of land and the electricity generation is considerable.'

But they don't mix well with homes. Precise regulations about proximity to dwellings vary according to turbine size, but big ones tend to be at least 500 metres from houses. However, that doesn't mean they are banned from built-up areas. I meet Dr Charles Gamble at a turbine site somewhat different from a blasted hillside: an industrial development zone beside Bristol Docks. He is co-founder of local energy company Community Power, whose latest generator has just started spinning above my head. It's one of the biggest on shore in Britain and is owned by a neighbourhood nearby. It's not a wealthy area but the community owners hope their windmill will deliver pennies from heaven. The ground around the base and approach to the turbine is just rough hardcore right now but there are competing plans: the local authority wants a solar farm but the community owners want an 'energy learning zone' to spread the word about green electricity. As such a building would not be a dwelling, it is permitted to be sited closer to a turbine. A few miles away a big warehouse and wine bottling plant has a turbine in the carpark, with trucks bustling round the base as the electricity keeps the production line going. Charles expects to see many more turbines in industrial zones: 'The ability of wind energy to coexist with other activities is a significant advantage.' Suburbs or business parks may not be the windiest locations but the grid connections tend to be easier and there are plenty of local power customers.

Charles is a pioneer of wind energy; his experience dates back to the first wind farms in the American West in the early 1980s. 'We installed them across the golden hills of California,' he tells me. 'Thousands of turbines, much smaller back then, across the ranches. The cattlemen below continued with business as usual except with extra income. Later, in the wheat prairies of Texas, the farmers thanked us turbine guys for putting in better roads. Landowners with marginal incomes from agriculture find wind very attractive.'

He makes a point amplified and applied in the UK by many backers of climate-friendly power sources: our uplands are blessed by strong winds and cursed by poor fertility. Wind farms have, well, the wind behind them whereas farming is an uphill struggle. It doesn't have to be one or the other; we can let one support the other.

But what about public opinion and political will? As wind farms are so visible, everyone has a view of them and everyone has a view on them. Data suggests that wind farms enjoy the support of 75–80 per cent of the UK population, with even two thirds backing them near where they live. There are similar figures coming out of continental Europe. Those who are opposed often feel it very strongly: they feel it in the heart, whereas support is in the head. Local enthusiasm is often heightened by local advantage, like giving some wind-farm profits to the community overlooking the wind farm. Despite this data, at the time of writing, the UK government has maintained an almost complete ban on onshore wind farms in England. Given their low price, low land take and low carbon credentials, this is completely absurd. The devolved administrations of Scotland, Wales and Northern Ireland are more welcoming.

Biomass

By far the biggest and most historic land-based energy source is firewood. For millennia, trees, shrubs or even peat have been an accessible energy source. Globally, it provides domestic energy for more than half the world's population and delivers more power than nuclear. Its use promotes deliberate firewood plantations, especially in developing countries, yet it is also a big driver of deforestation where planted supplies are insufficient. In the richer world it's used for wood-burning stoves and some pellet heating systems but these are usually the by-product of another land use like amenity woodland or commercial timber forest. But in the UK one species is now grown specifically for energy: willow.

At the plant science hub at Rothamsted in Hertfordshire they have 1500 different varieties of willow. Their collection was established after the First World War when willow was classed as a strategic resource. Imagine worrying about 'willow security' like we hear about food or energy security today. Its whippy sprigs are light, strong and, with skill, can be formed into nearly any shape. It was, in effect, the plastic of its day. In the early twentieth century, many villages had their own willow patch and a competent weaver who could craft items out of wicker. But during the years of grinding trench warfare, demand exploded because so much was being transported: shells in wicker paniers, weapons in wicker crates, even carrier pigeons were moved in wicker baskets and wounded servicemen moved in willow stretchers and later willow chairs. Willow made robust, light and protective packaging before packaging was a word in common usage. By the Second World War, although its use in civilian life had declined, wicker made a comeback beneath parachutes. As paratroopers dropped behind enemy lines, their vital kit often floated down behind them in a willow basket: light in the aircraft and offering some protection on impact with the ground. Nowadays the structural uses of whippy stems are mainly in the craft sector, but it's pleasing to know that containers of willow timber are still shipped out to India and Pakistan to make cricket bats; apparently English willow is appreciated at creases worldwide. Actually, there is one wicker container proving increasingly attractive as an environmentally friendly option: the coffin.

Willow also yields the active ingredient in aspirin, salicylic acid, and the bark of the tree was recognised for pain relief more than 3000 years ago. Its medical properties, particularly for combatting cancer and even Alzheimer's disease, are being vigorously pursued at Rothamsted too.

But the main, if somewhat stuttering, driver for willow's land use today is to make fuel. It grows fast, is easily chopped and dried and pulverised into wood pellets. Some is burned in power stations to make electricity but most is used to heat farm buildings or maybe a farm business park. I say 'stuttering' because it is still difficult for it to compete commercially with other fuels and imported wood chip. Its fortunes have waxed and waned with available government subsidy. In 2023 there were only thought to be about 2000 hectares of willow for fuel in the UK. By comparison we grow about 4000 hectares of strawberries. At the time of writing, a new biofuel strategy is being pondered by the UK government.

William Macalpine is Rothamsted's willow expert, especially when it comes to breeding and growing. He wanders through a maze of different plots, tenderly inspecting the young trees like a shepherd might tend his flock. Different varieties are best suited to different uses and locations and, when it comes to specialist breeding, he thinks we are only just scratching the surface: 'If you think about wheat, that has had thousands of years of breeding to go from a spindly grass in the Middle East to the staple crop we know today. We have only been working on wheat breeding seriously since the 1970s and have only just finished genome sequencing all the samples in our collection.'

He argues that willow can be grown in areas of the farm that are not much use for food, especially close to watercourses, where soggier land is perfect for the willow and the trees can reduce pollution reaching the river. Willow plantations need very little fertiliser and, compared to most cropland, are pretty good for wildlife. They are coppiced in the first year, growing in multiple stems thereafter. They can then be harvested repeatedly, every two to four years, in what is known as 'short rotation coppice'. Specially designed machines slice and dice the stems as they work up the row. William has calculated that for every one unit of energy put in you get twenty out, but he admits that this 1:20 figure will worsen rapidly if you have to transport the willow long distances in diesel-guzzling trucks as it is bulky, with a relatively poor energy-to-volume ratio.

William's willow breeding programme has had a striking success: in recent years they have doubled the yield of some varieties from ten to twenty tonnes per hectare, which would make it by far the best energy crop in the table above. I point out that it's still some way below solar, and William replies: 'You've got to look at biomass fuel rather differently as, unlike solar, you can store it for when you want it. In terms of efficiency, it stacks up pretty well if you are using it for heat, especially if you are replacing heating oil.'

I can see the logic for willow heating to warm a massive country house or a rural business park on an estate with some appropriate 'good for willow, bad for food' land. Yet this seems like quite a marginal case, especially when, in order to be economic, you probably need a big field, not a winding riparian ribbon, and conservationists might argue that biodiversity or the climate would be better served if that strip was deliberately managed to store carbon or shelter wildlife, instead of those outcomes left to be a happy accident of willow.

But William has a very credible and influential ally: the UK Climate Change Committee – the government's official yet independent advisors on how to meet our net-zero target. In their 'Sixth Carbon Budget', published in December 2020, the section on land use recommends: 'Planting perennial energy crops alongside short rotation forestry needs to accelerate quickly to at least 30,000 hectares a year by 2035, so that 700,000 hectares are planted by 2050. This could sequester 2 $MtCO_2e$ by 2035 and over 6 $MtCO_2e$ by 2050. When used with Carbon Capture and Storage (CCS) technologies this could displace a further 3 $MtCO_2e$ of GHG emissions elsewhere in the economy by 2035, increasing to 10 $MtCO_2e$ by 2050'.[*]

An area of 700,000 hectares is big – about the size of the county of Devon. The idea of combining it with CCS is that willow (or other energy crops) will suck the CO_2 from the air as they grow, and energy will be generated when they burn, but then the post-combustion CO_2 will

---

[*] $MtCO_2e$ – or 'million tonnes of CO_2 equivalent' – is the standard metric for quantifying the potential of a technology to cut greenhouse gases.

be captured and pumped into permanent storage, probably under the North Sea. The even longer acronym for this is BECCS – BioEnergy with Carbon Capture and Storage – and it crops up regularly in internationally recognised plans to slow climate change. There are plenty of arguments over if and when CCS will ever work at scale but, if it does, the most promising areas in the UK are in Merseyside and industrial clusters close to the North Sea like Aberdeen, Teesside or Humberside. Given the fact that using willow as a fuel only makes sense if you are not carrying it far, this raises the possibility of huge areas of farmland close to these CCS plants being planted with energy crops. One of the most tempting could be North Lincolnshire on the southern edge of the Humber, but it has fine agricultural land and a lot of food is grown there so, once again, when you consider land use and what might be lost or shifted elsewhere, I have reservations.

Biofuel

My love affair with biofuel began when I was seduced by a Swedish model. The model in question was a Saab 9-5 and the object of my affection lay under the bonnet: an engine that ran largely on plants in the form of bioethanol, a plant-based liquid fuel and a close chemical doppelgänger for petrol. It was 2006 and I was in Sweden to test-drive Saab's 'BioPower' creation for the BBC Radio 4 environment series *Costing the Earth*. Being a big kid and sharing the widespread infection of the 'broom, broom' disease (as my mother called it), I knew what was required: a 0–60 mph acceleration test. There is no law against G-force so I could floor it on the open road. The needle hit 60 after a shade over six seconds and, through a big grin, I claimed to have experienced 'guilt-free' driving. Plenty of others were also consumed by the passion; as he took delivery of his new Saab 9-5 BioPower, entrepreneur and businessman Sir Richard Branson said, 'I am convinced that biofuels are the way forward, both for the car and aviation industries, which is a vision we share with Saab.' The Swedes insisted they would soon be making the fuel from wood, even waste wood, and Sweden has plenty of trees.

Not long after that, I was again swept off my feet by another foreign suitor – this one sporting a plaid shirt and a baseball cap and dwelling in the American Mid-West. My chaperone was the then deputy president of the National Farmers' Union, Peter Kendall, and I was reporting on his delegation of UK farmers seeing what they could learn from the US bioethanol business. In America, then and now, a huge amount of ethanol for transport fuel is made from maize (corn in the USA). It is sold at the pumps as 'E85' and boasts up to 85 per cent bioethanol, therefore just 15 per cent fossil fuel gasoline, which is surely a climate win? My passion for bioenergy was underpinned by an appealing logic: any carbon emitted when the fuel is burned to propel a vehicle must have been absorbed from the atmosphere via photosynthesis in the first place. The more biofuel the less fossil fuel, so three cheers, man-hugs and top dollar for crops all round.

For a while, farmers, climate campaigners and car manufacturers all joined the love-in. In the early 2000s the European Union set a target for 5.75 per cent of fuel to be biofuel by 2010. The EU then went further with the Renewable Energy Directive, requiring 10 per cent of all energy in transport fuels to be produced from a renewable source by 2020. But, like so many a whirlwind romance, real-world truths soon began to nibble away at the fantasy. Repeated full life-cycle analysis studies of bioethanol in the US have shown it has not reduced global warming; indeed, a 2022 report says it has probably made it worse. This is because of the diesel used in all the tractors, the natural gas used in making the (liberally applied) fertilisers, and the nitrogen oxide – a gas with a global-warming potential 300 times that of CO_2 for every molecule – that those fertilisers themselves emit once on the field. And, of course, biofuels have to grow somewhere: either on previously uncultivated wild land or existing farmland, thus pushing that farming pressure elsewhere.

The accusation of biofuels driving deforestation was especially credible in Europe. We burn much more diesel than Americans as their tighter urban air-pollution laws have largely ruled it out for cars. It was trying to get round these US regulations that landed Volkswagen in so much trouble in 2015 when they were found to have cheated emission tests. Unfortunately, for great apes and other jungle dwellers, palm oil is a very popular ingredient of biodiesel and so chunks of their rainforest home in South East Asia were replaced by palm plantations in order that we could feel better about driving. A 2021 study for the clean-motoring pressure group Transport and Environment suggested that an area the size of the Netherlands had been deforested, destroying 10 per cent of orangutan habitat with overall three times the global-warming impact of their fossil-fuel equivalents. Soy oil, a potential ingredient for both biodiesel and bioethanol, is also largely of tropical origin and demand for it has been a critical driver of deforestation. It is true that rapeseed oil, much of it homegrown, has been the single biggest feedstock for biodiesel, and discarded cooking oils are now used too, but the proportion of imported oils seems to be growing.

I do regret my cheerleading for biofuels, but it was a brief encounter: my feelings went from sceptical to stand-offish to pretty hostile. But the policies that underpin the demand for biofuel have proved much harder to shrug off. The European Union still has its 10 per cent mandate for biofuels, and in the US most gasoline still has at least 10 per cent bioethanol (frequently 15 per cent or even 85 per cent in the Mid-West). Farmers in both the US and Europe are very fond of the extra demand for what they grow as such competition helps keep prices high, and the recent war in Ukraine, with its resulting squeeze on oil and gas, has added an energy-security claim to their cause.

So what does Peter Kendall think now? It's more than a decade and a half since we shared a few beers with friendly farmhands in Monroe, Wisconsin. In that time he has spent eight years as president of the National Farmers' Union, became a 'Sir' and developed his interest in climate change. He grows crops and rears chickens on his family farm in Hertfordshire, and I ask if he would grow crops for biofuels. 'No,' he replies. 'I don't like the idea of using a primary crop from good-quality land as bioenergy. In my pyramid of concerns over land use, climate change is at the top and most biofuels worsen climate change.' So how does he feel about his earlier enthusiasm for them?

'At the time I was an office holder for the National Farmers' Union,' he says, 'and finding new opportunities for farming was part of the job. I saw biofuels as a new opportunity, a new market for UK growers at a time when the price of grain was low. I might have been naive but, given the importance of finding climate-change solutions, it's important to try things and a few wrong turns should be tolerated.'

He is also aware of the likely demand for biofuels in the aviation industry and thinks we should continue research into how waste products like straw and cellulosic plant fibre can be turned into liquid fuel. As a final flourish he plays the multifunctional ace card: 'When you harvest rapeseed only 40 per cent is oil. The rest makes an excellent and widely used animal feed.' This raises a possibility that, in a few years from now, both the fuel keeping you aloft and the in-flight meal could originate from the same patch of land (although, of course, not flying or eating meat could free up land for a much more climate- or nature-friendly use).

While I, Peter Kendall and many climate and environment experts might have lost our appetite for bio-based fuels, globally they are still a massive and growing player. It's estimated that around 4 per cent of all the world's farmland is used to grow biofuels, which equates to more than 10 per cent of cropland. The World Resources Institute (WRI), a global research organisation focused on the wise use of our critical resources, has calculated that providing just 10 per cent of the world's liquid transportation fuel in the year 2050 from plant sources would require around one third of all the energy in a year's worth of crops.

Combined with other uses like textiles and pharmaceuticals, the role of farmland as a place to produce industrial feed stocks seems to be growing. Edward Davey, the WRI's Policy and International Engagement Director of the Food and Land Use Coalition, is very anti biofuels as they deliver scarcely any climate benefit, drive the loss of ecosystems and compete for land with food: 'At a time when more people around the world face hunger, the world's cropland should be used to grow food – not fuel,' he says.

So how does solar compare with biofuels? It varies, according to the sunshine hours on your solar farm, between 250 MWh and 700 MWh (megawatt-hours) per hectare. That is between five times and sixty times more power per field, hectare, square metre or whatever area measurement you choose than biofuel. Other studies go further still and suggest that on three quarters of the world's land solar PV would deliver at least 100 times more useable energy than a plant-based fuel planted there. The WRI rolls in the greater efficiency of EV motors to calculate that 300 hectares of corn (maize) for ethanol would propel you as far as one hectare of solar panels.

A further consideration is that solar farms don't need other rare commodities prized by farmers: fertile soils and good rainfall. One of the crop-based sources of energy that has grown most rapidly in the UK is anaerobic digestion. As the name suggests, these are like stomachs (but the size of a house), which use bacteria to break down organic matter and emit gas. This gas – mainly methane – can be burned to make electricity or used directly as a fuel. Originally promoted, quite sensibly, as a way of turning waste products like farmyard slurry or unwanted food into energy, it has become dominated by the consumption of maize grown deliberately as a feedstock: not a waste product at all but a farmed crop fed into an industrial belly. Maize growth often comes with the damaging side effect of increased soil erosion as the widely spaced plants leave a lot of bare ground and the late harvest when the dirt is wet and soft can harm soil structure. And, once again, is a terribly inefficient way of generating wattage, delivering around 20 MWh per hectare. The figures speak volumes: if you want a farm to make power, make it a solar farm.

While our future world will pivot towards electricity and away from fuel to burn, it is true that some industrial and transport users (notably aircraft) are going to depend on liquid fuels for decades. But electricity can make this too by creating hydrogen from electrolysis of water and combining that with waste CO_2, resulting in a synthetic liquid fuel.

So, given that the logic of physics, land-use inefficiency and climate change should all steer us away from growing plants for energy, why are we going in the other direction? One clue comes from another interview I conducted in the first flush of biofuel passion. In 2008 I secured an interview with the director-general of the United Nations' Food and Agriculture Organization, the body charged with improving global nutrition and food resilience alongside providing emergency assistance. Jacques Diouf received us warmly at their Rome headquarters and I asked how he saw the growth of biofuels worldwide, fully expecting him to condemn them as a threat to food supply and a contributor to hunger. How wrong I was. Instead, he welcomed them as another market for farmers, as something that could bring investment into the sector, drive up prices for farmers, discourage them from leaving the land and encourage agricultural production generally. In essence, these are the same motivations as Peter Kendall's during his Mid-West farming visit to the powerful agri-sector lobby groups in the US and EU: biofuels provide a great income for us. So while the prices farmers get for food don't compete and the voice of displaced wildlife is so feeble, how can biofuels be shifted from their throne?

The WRI argues that biofuels are a waste of land (the only exception being when the fuel is a by-product of another activity like wood waste from the timber industry or residues from food production). It has two principal criticisms based on land and climate. As regards land, the bioenergy targets from countries around the world would either gobble up more land or divert much of the food crop into fuel tanks. The WRI points out that many govern-ments have targets for 10 per cent of transport fuels to come from plants and, if that were followed globally, by 2050 it would demand 30 per cent of all the energy in crops grown today: bad news for wildlife, human life or both. For millennia we powered our civilisation on photosynthesis, be it wood, olive oil or oats for horses, but now we enjoy lives based on what lies beneath: the gigantic energy of coal, oil and gas. If we used all the biomass harvested today – food crops, grasses, residues and trees – it could provide just one fifth of the projected energy demand by 2050. Plants simply cannot keep us fed and fuel our machines.

The WRI's climate argument has two prongs. The first is that the climate impact of fertilisers, energy and chemical inputs to grow the crop or tree are often ignored. The second is what they see as an 'accounting error', whereby land used to grow the energy is seen as neutral but in reality is usually doing something else like locking up carbon, providing habitat or yielding food. Given the absence of almost any redeeming features for nearly all biofuel used on a large scale, the WRI says we must end targets and tax subsidies for biofuels while specifically excluding bioenergy produced on any dedicated land from renewable energy standards.

Very low-carbon energy generation is the single most effective way to slow climate change and, deployed smartly, none of the dominant technologies will have an outsized impact on land use. Harnessing enough power for the future from wind, sun and even the atom can be done without seriously threatening food production or space for nature. Solar clearly needs the most thought as, by definition, it needs area to function. But, as we've seen, it loves to multi-task, combining surprisingly well with farming, housing and businesses and it even works on water. By contrast, biofuels are the bad apple: not only are they very rarely low carbon, but also their land take is already massive and predicted to grow. While there are some very specific places and uses where fuels crops can be considered good land use, generally biofuels are a dumb use of land, eroding the natural world and human food supply.

4

FARMING, PART ONE:
ARABLE

J ust south of the Scottish border, overlooking the
North Sea, lives a champion. Having shunted a New
Zealander out of the *Guinness Book of Records* to take
his place in 2015, Rod Smith's wheat yields are unbeaten:
he can harvest 16.52 tonnes per hectare. That means an
area the size of a football pitch on his farm could provide
366 British citizens with bread for a year. The average UK
yield is half Rod's, at 8 tonnes per hectare. Globally the
average is half that again, around 4, which would feed only
91.5 people. Rod's best fields can feed four times as many
mouths. Alternatively, if everyone farmed like Rod, we
could give three-quarters of our farmland back to nature or
give it over to housing or transport or energy or any of the
myriad land uses that concern this book. Clearly that bald
proposition is a bit absurd, as soils, weather, crop varieties
and farming skills all vary hugely, but the point is that yield
matters. We need to produce more with less.

Rod's farm is in Northumberland, overlooking the Holy
Island of Lindisfarne from where, in the seventh century,

St Aidan planted Christianity in northern England. I went there to interrogate Rod's seemingly miraculous farming and test my core suspicion: is he a chemical junkie who achieved massive crops through titanic fertiliser use? Before revealing Rod's response, we need to address another question: what is the beef with fertiliser? Let us start with the science behind it, the history and the upside before looking at the future and the downside.

Plants need nitrogen to grow. It is a key element in their cells, as it is in our own. Air is 78 per cent nitrogen (N_2), but plants can't just suck it out of the atmosphere, they need a middleman. Some plants – notably for farmers, the legume family of beans, peas and clover – pair up with bacteria in the soil called rhizobia, which can turn inert N_2 into biologically useful NH_3 (ammonia) in a process called nitrogen fixation. The bacteria live in pinhead-sized bobbles called nodules on the plants' roots and make the ammonia available for plant growth. Most farmed crops – especially the world's staple cereals like wheat, maize and rice – can't do this. Instead, they use mycorrhizal fungi, which can feed on rotting organic matter and sugars from plants. The barely visible fungal strands exist between or within plant roots and they trade their scavenged nutrients, like ammonia and phosphorus, for the carbohydrates that the plant has created through photosynthesis – a skill mushrooms lack. This deal between plants and fungi underpins much of life on Earth. Indeed, it was this fungal relationship that allowed the colonisation of the land by plants 460 million years ago, before plants even evolved roots.

Like all swaps, each side has to give up something, but we humans are only interested in harvesting one side of this equation: the plant crop. Fungal wellbeing matters nought. What if we could offer the cereals we crave a better deal: food for free, growth without sacrifice? Step up, seaweed, bird droppings and two German chemists.

Kelp, the big fast-growing seaweed common on much of the coast and islands of the UK, is rich in nitrogen, phosphorus and potassium – the NPK formula so beloved of gardeners. In the eighteenth and early nineteenth century a huge industry developed to supply it dried, or even as ash, to arable farms. But that was gradually edged out by an even richer source of these chemicals: seabird dung, or guano. Bird colonies on outcrops in the south Atlantic and especially South Pacific near Peru and Chile were piled high with thousands of years' worth of what left the back end of birds. Its potency as a fertiliser is such that it became known as 'white gold' and millions of tonnes were mined from these islands and shipped to Europe. But then came a more industrial answer: in 1905 the German chemist Fritz Haber succeeded in taking nitrogen from the air and combining it with hydrogen, under enormous heat and pressure, to create ammonia. His process was industrialised by Carl Bosch from the BASF company and the era of artificial fertiliser was born. Its use spread steadily across Europe, America and Asia as crops flourished under its reign.

With its use the plants are getting a great deal: they no longer have to give up sugar to get nitrogen. It's there, on tap, for free. In the second half of the twentieth century as human populations boomed, plant breeders realised they could develop varieties well suited to the use of chemical fertilisers. The stunning scientific success of the Green Revolution, which kept us fed by developing high-yielding varieties, relied heavily on the Haber–Bosch process and the ammonia it delivers. It is reasonable to say that around half of the world's population of 9 billion people are alive today thanks to artificial fertiliser. So where is the downside?

The manufacture and use of fertiliser emit huge volumes of greenhouse gas, the chemicals within it pollute our air and water and it undermines soil health itself. As described above, the principal ingredient in artificial fertiliser is NH_3 (ammonia). The H (hydrogen) needed to make it comes from breaking apart the gas molecule CH_4 (methane). This requires a lot of heat, but during the reaction the C (carbon) in CH_4 also joins up with oxygen to become CO_2 (carbon dioxide), so both the energy required to set off the reaction and the chemical reaction itself are sources of greenhouse gas. Combining the hydrogen and nitrogen to make the ammonia needs yet more energy. The next step in fertiliser's global-warming story comes when you spread it on the fields: much of it never reaches the plant but evaporates as nitrous oxide: a pollutant and gas with 300 times the warming effect per molecule of CO_2. Add all this together and fertiliser is responsible for around 3 per cent of all human-made climate change – a proportion similar to aviation.

Frustratingly, although fertiliser prices are at record highs at the time of writing, so much of it is wasted, escaping into watercourses and never reaching the target plants. Instead, it runs off into streams and lakes where it's toxic to much aquatic life but lapped up by algae. This can cause algae to multiply rapidly, giving us so-called 'algal blooms', which can kill other life by shading them, poisoning them or, when the algae itself rots, sucking all the oxygen out of the water and causing suffocation. Resulting 'dead zones' in wetlands and coastal waters are becoming all too common. Northern Ireland's Lough Neagh, the biggest lake in the UK by surface area, has been poisoned by blue-green algae mainly caused by agricultural run-off.

Furthermore, nitrates in drinking water can be damaging to human health, and water companies have to monitor concentrations to make sure they stay within safe levels. The combination of these risks means that in England there are 'nitrate vulnerable zones', where farmers are restricted on the amount of fertiliser they can use and on how they store and spread animal manure. It has even been shown that rainwater itself now contains increasing amounts of nitrate, showering diluted fertiliser on the ground and favouring plants that need it, such as stinging nettles, and disadvantaging those that don't, like many wildflowers.

Today, given what we now know, chemical fertiliser is in a very weird place: simultaneously keeping many of us alive while threatening our long-term survival. That is a big enough dilemma, but it's allied to another one core to this book: whether farming that is dependent on artificial fertiliser can really be considered intensive in the sense that its impact is confined to a small area. In a narrow sense, it can be, in that much food comes from a small patch. However, I would argue that the full life-cycle footprint – including energy demand, global warming and pollution – is massive, giving chemical fertiliser an *extensive* environmental impact. This does not make it a smart foundation for agriculture; it is not sustainable and cannot be relied upon in the long term.

So is champion farmer Rod Smith hooked on this short-term fix? It's the month of May and, taking a few steps into one of his fields, Rod picks an ear of wheat and rolls it between his fingers. 'Still green, green as leeks,' he remarks approvingly before counting the grains. Eleven of them down each of four sides. 'That should make around 9.5 tonnes a hectare. And that's just on the edge. In the middle, the harvest will be even greater.'

Yield really matters to Rod. He's a farmer who gains satisfaction, even perhaps validation, from the fact that he grows a lot of food for a hungry world, and he describes with great passion the moment he broke the record: 'The adjudicator sat with me in the combine harvester and we were recording a rate equivalent to 22 tonnes per hectare and he said, "I can't believe it, I've got to take a photograph of that." I said, "You've got plenty of time 'cos it'll be the same on the way back," and he said, "Don't be so stupid, it's a one-off." But it wasn't a one-off. We did 2–3 hectares of 22 tonnes per hectare. What I learned from that was what is possible, feasible.'

Not every part of the field grew at that phenomenal rate and Rod ended up with the average mentioned above of 16.52 tonnes per hectare. That's still a record-breaking figure, but was it all down to performance-enhancing chemicals? 'We did add a lot of nitrogen,' he tells me, 'but that year it was definitely soil health that won us the record, without a shadow of doubt, because of what the beans did.'

Immediately before planting the wheat, the field had been home to a failed crop of beans. All their plant matter – roots, stalks and those all-important nitrogen fixing nodules – were all ploughed back into the soil. Chemical assistance resting on top of very favourable soil biology delivered the medal. It was a high point for a farm in transition.

'Back in the end of the last century, when my father was in control,' Rod says, 'the soil smelled sour. There was no aeration, everything grown was taken off, I'd almost say they were raping the soil. Everything was very cheap: fuel, labour, fertiliser. And the price they were getting for their crops was huge. They had five full-time men plus a shepherd. I certainly couldn't afford to have that now. I'd be out of business. They were almost obsessed with getting the soil surface super-fine and would go over it two or three times. Unbeknownst to them, that was actually damaging the soil structure.'

Yields began to drop, weeds grew and so did bills for pesticides, herbicides and fertiliser. At the same time Rod discovered the 'plough pan', which is a layer of soil about 30 cm down, beyond the reach of the blades, that is rock-solid due to the weight of machinery regularly passing above. Water, roots and air can't get through it. One solution is to deploy more metal and diesel in the form of a subsoiler: a curved spike about 60 cm long dragged through the field. But Rod felt that tractor passes were the problem rather than the solution. The soil was solidifying under the weight of the machinery and, with no strong root systems to break it up, was also starved of life. The solution was cover crops: plants like mustard, radish or phacelia (a purple flowering plant that acts as a kind of green manure). Their vigorous roots open up pore spaces in the ground and, when the whole plant decays, it gives the soil a meal. Rod plants them immediately after harvesting his cash crop and, after growing for six months, in the spring they are either grazed by a neighbour's livestock or cut and left to rot.

The better soil structure saves money too as Rod can cultivate it in half the time with a lighter machine using a quarter of the fuel. The cover crops' value is so obvious, Rod can see no attraction in leaving ground brown.

'Why would you leave bare soil, with the amount of rainfall you can get up here in the wintertime – great thunderplumps? Where are all the nutrients going to go? Unless gravity laws have changed dramatically, it's only going one way and that is straight into the drainage system and out into the watercourses. Lost plant food for me and bad for the streams. So the sooner you get the cover crops into the ground, the sooner they'll suck up all those spare nutrients into those lovely crops and then you're going to break that down in the springtime, ready for your next farm crop. It just makes sense.' Used correctly, Rod says, living cover crops soak up excess fertiliser (which might otherwise leach into the watercourse), thus shrinking the fertiliser's pollution footprint. And the decaying cover crop delivers slow-release nutrition, making less fertiliser required in the first place.

Rod pampers his cover crops, giving them at least as much attention as his commercial varieties. But it's not his only new-found ally; the other lies beside the A1, the historic trunk road from London to Edinburgh that skirts his farm just 16 km (10 miles) short of the Scottish border.

He admires a brown pile about 50 metres long with a granular crust before showing me it has warmth like body heat just beneath the surface.

'It's a beautiful product. I've had the Environment Agency and Natural England [here] to show them. Everything in the periodic table is in there and it's teeming with life. I think this is the way forward for farming … revolutionising it,' he says, before adding, 'It's not just a pile of muck.'

Er, in a way it is. And all this hyperbole might sound a little overdone for what is essentially dung and straw. He gives his stubble from all his arable fields to a neighbour with 2500 cattle and it gets used as bedding, especially over winter. Rod then takes it back when the cattle have soiled it. No money changes hands. The lengthy mound of compost, called a windrow, is turned every ten days or so to keep the temperature high enough (around 55 degrees Celsius) to kill weed seeds and to even out the hot spots of decomposition and allow more oxygen in. The process also steadily dries and condenses the compost, meaning it's lighter and cheaper to spread. Rod reckons he can get the same nutrition from one tonne of his manure as typical farmers might get from 10 tonnes of wet manure. He also loves the fact that his decomposing dung heap is next to the main road: 'As we are right on the A1, a large amount of people pass and say, "what is he doing now?" That is a main reason why we are doing it here – advertising it to other farmers.'

Once spread on the field, the effect, alongside the benefit provided by the cover crops, is transformative. It leads to richer soils with more fungi and microbial life and a better structure, which increases the resilience to drought or flood.

A key measure of the soil health is the amount of organic matter, or soil carbon, it contains. Increasing this is good for both crops and the climate as more carbon captured in the soil means less floating free in the air. The heavy use of ploughing and chemicals in the past meant Rod was starting from a very low base when it came to soil carbon. He tells me it was down to about 1 per cent but is now up to about 3–4 per cent in five years, and he wants to get it up to 6–7 per cent. 'I'll never stop using compost and cover crops,' he says. He would like to get the local university in Newcastle along to do a proper independent measurement of what that means for total carbon absorption across the whole farm. It could be a sizeable figure and, though climate change isn't his key concern, would be a very welcome fringe benefit. But I want to know if his chemical fertiliser use has dropped.

'We've reduced our nitrogen fertiliser over the last few years by 30 per cent on the spring barley,' he replies, 'and look – the crops are so much greener, lusher and thicker. This barley – sown at the end of March after a cover crop and compost, no ploughing – it's looking absolutely superb.'

However, its use on wheat is only down around 10 per cent because Rod doesn't yet feel confident enough of his compost, cover crop and low tillage recipe to be sure of keeping yields up. He wants to cut his costs not his harvest. He hasn't spread any artificial potassium and phosphorus because all that comes from his homebrew compost. Food production, coupled with steadily shrinking environmental damage, is the priority on Rod's farm and his fields are dedicated to that goal. But what about nature – does that get a look-in?

It isn't ignored, just separated – often by merely a few metres. He has widened his grass strips, planted pollinator plots with sunflowers, linseed and ox-eye daisy; his hedge-rows have broadened too. There is even a large pond in the middle of the field that delivers both habitat and a sink trap to prevent excess nutrients and soil being washed into watercourses. During half a day on the farm, I see brown hares, grey partridge scuttling about and buzzards mewing overhead. Rod has even given up some land to the North Sea by deliberately breaching the defences and letting the tide flood in to create a brackish wetland. Wading birds love it. Across the whole property about 10 per cent of the land has been given back to nature, but the wild animal that really fires him up is the one that shares the field and likes his cultivation: the gardener's friend, worms. They even dictate his farming schedule.

'We don't cultivate in darkness 'cos I have found that the biggest worms only come up at night. The really, really big ones, the big drainers only come up at night and you don't want to be killing them. They take the foliage from the surface and take it back down, and the more worms you've got the more they are pushing the soil through themselves. And we've got worms right throughout our soil profile. The other day we pulled a very shallow cultivator, about 5 cm deep, through the field and the soil behind was thick with gulls eating the worms. We never had that before, but the worms are now in abundance from the surface downwards.'

For Rod, healthy soil is as close as he can get to an effec-tive insurance policy at a time of climate change. One of his mantras is 'I'm always trying to reduce the risk'. He has seen first-hand how the weather is getting more extreme and 'blockier' with longer stretches of testing conditions. Lifeless, compacted soil can be coaxed into delivering a harvest with sufficient chemicals, but those chemicals won't protect plants from drought or flood. Healthy soil is so much more resilient – holding the right amount of water when it's dry and draining when it's wet.

Just as my time with Rod draws to a close, he reveals possibly the most significant change on the horizon: he's going to become a mixed farm. His son, Will, is just finishing agricultural college and will start a herd on his return. It's hard to overstate the real and symbolic importance of re-combining animals with arable. Many people chart the start of damaging monoculture and plummeting soil quality to the time when crops and cows parted company. Across much of the richer world in the latter half of the twentieth century easy access to chemical inputs and global markets meant farmers specialised in fewer products. Taking a cue from industry, efficiencies could be achieved by narrowing and heightening the skills of staff and the design of machinery. Many countries saw their land divided into arable or livestock zones. In the UK the animals went west to the wetter, hillier parts and the crops went east. But now there is too much manure in the hills and too little organic matter in the east. Pretty much all natural environments contain plants *and* animals; the most vibrant ones nurture plenty of both. Rod is really excited about the change. He sees animals as a new asset for farming.

'I've been fencing a field or two every year, which for an arable unit is really unheard of. By the time [Will] gets home the whole farm will be done. Having all the fields fenced will give us total control over how and where to use the livestock. For instance, I referred earlier to animals grazing off the cover crops, but sometimes the stockmen keep them on there too long in the wet and the soil can get compacted. My son and I can work together well with complementary not competitive enterprises, and it fits so well with the regenerative farming idea.'

The wheat-yield champion welcoming cattle in the yard – that feels like a smart turn.

More and more farmers, like Rod, see respect for soil biology as the key to long-term viability. But this idea that you really can have high yield and low environmental impact – known by some as sustainable intensification – is so contested that many think it is wishful thinking. It would be fair to point out that Rod is still using a sizeable, though shrinking, tonnage of chemical inputs, so I want to meet another farmer who appears even closer to making that wish come true.

Once I've rolled to a halt, the gate to Brewood Park farm opens automatically after barely a five-second wait. The drive is lined with blossoming cherry trees above sporadic cow parsley, and beyond are what appear to be conventional arable fields. The farmyard is bounded by pleasantly aged red-brick barns and a stout Victorian dwelling. The farmer himself climbs down from a tele-handler and greets me in well-worn overalls badged with a make of tractor. His handshake is firm, dry and rough. So far, so normal. But Tim Parton's head and hands are delivering something quite revolutionary: 'I'm increasing biodiversity: pollinators, moths, spiders, birds,' he tells me. 'I'm still producing the same if not better yields. I'm putting carbon back into the soil. The whole farm is alive and the farm profitability is up. We are creating soil for generations to come and business for generations to come. There isn't anything not to like.' This is the kind of win–win that sets off my alarm bells of journalistic scepticism. Surely it is too good to be true.

Tim farms 300 hectares just north of Wolverhampton in the heart of England, growing wheat, rapeseed, barley, spring beans, oats and flower crops like cornflowers, marigolds and lupins. He's worked here his whole life, taking over from his father as farm manager, and says: 'I don't use fungicides. I don't use insecticides. Every field has a cover crop. The soil is almost never bare. I farm with the heartbeat of nature.'

Much of his language is unusual for a commercial arable crop producer, as he is aware. He continues:

'The whole [natural] system was working perfectly well until we came along and we just made a mess of it. We are arrogant. We think we know better than nature and interfere with all these chemical poisons. I sound like a blimmin' sandal-wearing hippy, but I've come full circle. I used to be that person putting chemicals on, but then you see the side effects and the damage it does. We all want a fungal soil, so why the hell would you put a fungicide on? The clue is in the title.'

But it wasn't harm on the farm that triggered his conversion, he tells me:

'The big change for me was when I went through a period of anxiety and depression and my wife, who worked in mental health, didn't want me to go on drugs but get my nutrition right. So I got tested and corrected all the imbalances in my diet and got better. And then I thought, "Well, if it works for me, surely I can do that for my plants." So I read a lot of books on plant health and plant nutrition. I keep disease out now by knowing what elements to put in [the soil]. For instance, potassium helps the plant resist the damaging fungus *Septoria*.'

In essence, Tim believes that crops are more resilient with the right nutrition, pest attacks are minimised by plenty of predators, and vigorous growth is best achieved by improving relations with fungus and soil microbes – nature's way of providing that all-important nitrogen and phosphorus to the plant. I said earlier that humanity wasn't very interested in helping soil fungus get a good deal from plants because we eat the plants and so we want them big and selfish, but Tim believes that is a fight with the natural order, delivering a brief victory followed by crushing long-term defeat.

'Soils are in such a bad state across the country,' he says. 'It's not unusual for me to go onto a farm, digging as I go round and find one worm. Every time you plough you've halved the worms. If you are doing that year after year, it is impossible for that worm ecosystem to get established.'

And he has particular disdain for farmers whose bare ground or poorly managed drainage allows the soil to wash away and be lost altogether: 'I've always had a passion for soil and could never understand why we abused it so much.

I've seen so many farmers laugh at the fact that the river has gone brown and they know they have caused it. It's not funny. You can't go and shovel that soil back up, it's gone. It's clogging all our estuaries up. It takes so many thousands of years to make and you are just letting it flow down the river. It's finished.'

Changing how he nurtured both himself and his farm did prove a win–win but, for many years, relations off the farm became tricky: 'The early days were very lonely and challenged me mentally because I'd go to conventional farming events and people would be very pleasant and say, "How are you?" But they wouldn't talk farming. It's as if it scared them. Or they said I was mad and it wouldn't work. It's isolating and I had loads of self-doubt.'

Even his news intake challenged his decision to abandon the chemical weapons: 'I had to stop reading media as there were always scare stories in things like the *Farmers Weekly*, saying this is going to be the worst year for *Septoria* or brown rust [fungal problems on cereals]. If you read it, it gets into your head and you start panicking.'

Given his respect for soil, the life within it and a hostility to chemicals, I wonder why he didn't become an organic grower – after all, the key accreditation body is called the Soil Association – but he says: 'I am not really an idealist, and the problem with the Soil Association is that they won't let you use lots of the natural nutrition that I use. I call what I'm doing beyond organics. I'm not using poisons, I'm just using natural elements to keep that plant in balance. But sometimes everything can't come from the soil and you have to step in.'

There is one poison he does use, albeit sparingly, and believes is totally justified by the outcomes and the punishing nature of the alternative: glyphosate, the commonest weedkiller in the world. Like many regenerative farmers (including Rod Smith), he plants cover crops after his commercial harvest to protect the soil and keep reaping the benefits of photosynthesis. But, when the time comes to prepare the field for the main crop, the covering plants need to be killed off. That time is winter, and Tim loves an enduring frost as it means the field is so hard that he can get heavy machinery on it without compacting the soil.

'Minus four or below means I can use the roller,' he says. 'But some years we don't get frosts and this is when I resort to weedkillers instead of ploughing [which would be another way of killing off the cover crop]. I believe that to be the better of the two evils!'

Tim is convinced that limited herbicide application is much less harmful than ploughing to his fungal friends in the soil. But while ploughing is permitted by the Soil Association, glyphosate is not – another reason why Tim couldn't roll with organic orthodoxy: 'They allow you to stop weeds by cultivation – ploughing or hoeing – but fungi are so delicate. You put a plough through the earth, it's like someone putting a ball and chain through your house.'

Tim is nothing if not inventive, and he takes me out to see his mechanical and chemical creations. Our first stop is for a massive multi-tool towed behind a tractor, which he tells me wasn't off the shelf: 'The first thing we did was take an angle grinder to it and then welded on that entire back platform with those tanks and hoppers.' It looks like a crazy steam-punk contraption with fat tyres, slicing discs, saw-toothed cogs, hoppers, pipes and nozzles. It rakes, plants, feeds and fertilises all in one pass. Cutting traffic on the field is critical to reducing fuel use and soil compaction. He credits his friend and farming engineer Trevor Tappin for this unique creation before proudly showing me a roller they co-designed. Two hefty cylinders, each about a metre wide, are adorned with regular raised metal slats. It's for breaking the stems of cover crops or weeds, thereby killing them without chemicals: 'This goes on the front of the tractor and each one floats according to the shape of the ground.' I suggest it works like the electric shaver you see advertised that moulds to the contours of your face, and Tim replies: 'Exactly, they undulate to follow the ground. We've got air pillows so I can set this to go across your foot and tickle it or set it to crush every bone. I've got complete control to get the perfect trim. The point is to damage the [unwanted] plant but not the soil.'

The best time to use such tools can be very precise, he says: 'If it is frosty, I'll get up at three o'clock in the morning, because the freeze means the tractor won't damage the soil, and go and roll the cover crop, which will kill 95 per cent of it. All that's left behind is a "thatched soil" of dying stems and leaves that help retain moisture and reduce compaction from heavy rain.'

As we walk across his barn, the gathering scent of fermentation suggests we are approaching his homebrew. There are two plastic tanks, a cubic metre each. One is filled with a coffee-coloured liquid that smells like a mix of beer and marmite; the other is empty, save for a thin foamy residue ('bio-film', I'm told). In one tank Tim is cooking up a liquid to prevent unwanted fungal attacks. He buys in bacteria developed in a clean lab to fix nitrogen from the atmosphere and release phosphorus, along with many other elements. Water is added and the brew is maintained at 18 degrees Celsius and aerated for three to four days, which Tim says 'multiplies the bacteria and keeps the cost down. I can do an application for 80 pence per hectare, whereas some farmers are paying £65 per hectare for the same result.' The other tank is for fertiliser, where he is marinating manure-rich compost in warm water with an 'air pillow' at the bottom of the tank. That pushes a huge amount of oxygen through the liquid, forcing microbes off the organic matter and allowing them to reproduce.

It's not just the creation of the fertiliser that is unusual, though, it's also the way he applies it – from above not below.

'Foliar feeds are five times more efficient than soil applied. I apply through a conventional sprayer normally at night when temps are lower. This enables me to use far less and still get high yields [profit is and always will be king]. And this technique cuts my carbon footprint and I'm not polluting our valuable planet as I am only using a quarter of the nitrogen that I used to use. I don't apply potassium or phosphorus fertiliser as good soil will provide, but if the weather goes too cold or dry I will step in to foliar feed any element which is deficient to keep balance.'

Tim's techniques have given him stunning vital statistics. In a land scarce world, they make him a model farmer.

'My best yield last year of wheat was 12 tonnes per hectare with 50 kg of nitrogen. Most farmers would be using 250–300 kg. We did some trial plots and we didn't get any more yield where I was using 250 kg per hectare of fertiliser than where I was using 50 kg and foliar feeders. It just shows that all that nitrogen that I put on years ago was a total waste of time. Every hectare has got 10,000 tonnes sat above it in the air; it's just a question of converting it [to be available to plants].'

The important measure is how much his chemical fertiliser has decreased overall, as it is the manufacture and use of artificial nitrogen that has such a large impact on global warming.

'At the start of my regenerative agriculture journey, thirteen years ago,' he tells me, 'I was using an industry-standard 250 kg of nitrogen per hectare to grow 9.5 tonnes of winter wheat, on average. Last season, I had managed to reduce total nitrogen application to 74 kg per hectare on average without sacrificing yield. This represents a 70 per cent reduction in nitrogen dependency.'

It's hard to overstate the importance of this. It represents a slashing of chemical fertiliser use and all its associated damage to water and the climate, but Tim rejects one commonly used description of his type of farming: 'The minute you say low-input farming, people always get this vision of a farm that's untidy, that's not getting the maximum yield out of the crop and they are just farming for fun. I didn't want to be branded with that because I work extremely hard to get these crops to their fullest potential. I'm just making an informed decision because I have got all the information coming in. I don't mind spending the money if it needs spending, but I would sooner not spend the money and still get the maximum yield.'

He likes to call it intelligent farming (though he is not implying that the rest of farming is dumb). I would also call it creative. He is doing so much himself – designing machinery, making fertiliser, measuring everything.

'I call it intelligent farming because we've got all the information we need to farm in a sustainable way, that will regenerate the farm for the planet, and we've got something for generations to come. I know we've got a big population but I am proof that we can produce the same amount of food, if not more, and heal the planet along the way. We cannot use food as an excuse to pollute or there will be no people and a rotten planet.'

And he is no longer a lost prophet: 'We had a farm walk here in 2012 and nobody was interested. Now my diary is packed with so many visits and talks, I struggle to find time to do the farming... but I love it.' His shelf groans with British farming awards too: Soil Farmer of the Year, Arable Innovator of the Year, Sustainable Farmer of the Year, Innovation Farmer of the Year.

Whether growing food for us or feed for livestock, arable farming is the bedrock of global nutrition. It's what allowed us to quit hunter-gathering and develop modern civilisation. Although its recent environmental impact is punishing, Rod, Tim and a growing band of enlightened land managers are turning that round towards becoming rewarding.

5

FARMING, PART TWO:
LIVESTOCK

Two thirds of the farmed area of the UK grows grass not crops. It is grazed by cows, sheep and wild deer. The worldwide proportion is similar. What are the smart options here?

I grew up in the breadbaskets of Cambridgeshire but spent most holidays in the Scottish Highlands. My parents, John and Peg, and my aunt Janet bought a house on the Isle of Mull in 1965. I say house, it was actually a ruin with sheep dung about half a metre thick on the floor. Over the next decade, it was gradually restored and made a holiday destination for two or three trips every year. My dad was a keen mountaineer and encouraged me and my two older sisters to get out and enjoy the Highlands in all weathers. Through both culture and personal experience, I nurtured a deep affection for the rugged beauty of the place and assumed, without really questioning, that this stripped-back geography was its natural state. And then I went kayaking up Loch Shiel with two of my own sons.

Loch Shiel is a long, thin, inland lake that stretches

south-west from Glenfinnan. A point just west of Fort William now famous for the 'Harry Potter Viaduct', as the Hogwarts Express is frequently seen steaming across it in the film, and nearby carparks and The Viaduct View Cafe are regularly seen crawling with Potter-heads in the real world. Look the other way and Loch Shiel itself has its own movie moment as the place where 'Highlander' learns he is immortal after being tipped out of a small rowing boat by Sean Connery in the 1986 film of the same name. Movie makers are clearly with me in a love of this landscape – maybe I've left out a key land use from the book: film locations!

It was a three-night trip and we had decided to wild camp, stopping when fatigue, low light or the need to feed got the better of us. One evening, we paddled up to an island about the size of a tennis court with some rounded rocky margins. But beyond these whaleback boulders was a lush jungle of tall, varied trees and an undergrowth of fleshy plants between saplings. It reminded me of another film set far from here, a fragment of Jurassic Park. It was the sheer rich abundance of the place that struck me, especially in comparison to the barren shores of the loch. It was a cornucopia bursting through the water – a foreground Eden with a bleak backdrop. I knew that the western fringes of Scotland could be lush, with some famous gardens holding an extraordinary variety of plants, but those are tended and managed environments. What was it that made my tiny speck of a campsite so lively?

There were no grazing animals – no deer, no sheep, no cattle – and just maybe an occasional, ambitious, migratory goose, but I have my doubts. Now awoken, I saw this simple truth everywhere in our uplands: where the herbivores ain't, plants show no restraint. Driving through the hills, I noticed a glen divided by a deer fence. On one side there was just a mere coating of nature cloaking the bedrock; on the other side, where deer were excluded, shrubs and scrub and trees covered the slopes with a thick green fur. Even at some altitude, like on the islands in the lochans of Rannoch Moor at above 300 metres, trees flourished. It is well known that most of the uplands of the British Isles were wooded thousands of years ago, but it was these glimpsed fragments of what it could be like – and once was like – that really opened my eyes, and I had a sudden realisation of how lively the surrounding landscape can be.

Shortly afterwards, I visited Glenfeshie, one of the Scottish estates where the landowners are committed to rewilding (although they are sensitive to the toxicity and partial inaccuracy of the term). There's no sheep farming and the deer population is controlled with guns, not fences. They argue that shooting is more like the natural action of predators than a wire barrier that also stops the movement of other terrestrial creatures and even low-flying birds. At that time, there wasn't an instant change as you entered their land but more a growing sensation that something was different. Although I had become accustomed to the British uplands, this felt wilder and more jumbled up. Most hillscapes consist of land for grazing, bogs, conifer plantations or mature woodland, each generally occurring in defined blocks. But this was a glorious mess of tall grasses, untamed rivers, damp patches, scrubby trees and taller relatives. In short, it felt like the wild.

A hill farmer reading the paragraphs above might well be rolling her or his eyes while thinking, with some justification, that I see with the eyes of a tourist – a visitor who neither lives there nor makes a living from the uplands. So let me spell something out: I am not an advocate of wholesale rewilding of upland Britain, but I do ask whether much hill farming could be done in a smarter way. We are talking about a huge area – about 40 per cent of the UK – so it's right to ask whether the traditional land-management techniques are delivering enough food, or wildlife, or carbon storage, or renewable energy, or flood-water storage as to be unquestionable? I think not.

Rachael and Geraint Madeley Davies question everything in the often conservative world of Welsh hill farming. They own and rent land on the eastern edge of Snowdonia National Park, not far outside the county town of Bala. Spread over three locations, 10 km (6 miles) apart, the farm starts from about 229 metres and goes up to 670 metres at its highest point. The total area is about 490 hectares, including 14 hectares of ancient oak woodland. It's 85 per cent SSSI land (Site of Special Scientific Interest – a legal conservation designation in the UK for places of high natural or geological value), so there are regulations on what they can and cannot do. And it is 85 per cent organic.

Geraint was born into farming but takes little of what his family taught him for granted. Rachael grew up close by but became a London-based barrister specialising in employment and dispute resolution, before Geraint ('inconveniently', she jokes) proposed. She moved to work as a commercial manager for retailer Marks & Spencer, ran a farm shop and online sales for big Welsh estates. Now, she mixes on-farm work with agricultural consultancy on the tricky task of decarbonising dairy and livestock farming while also making them more wildlife-friendly. We have a chat in their kitchen before heading out to the hill, and I ask them what they think the farm is for.

'I think our farm is very much for food production,' Rachael explains, 'but we're also delivering for the environment and we really strongly believe that they're not mutually exclusive, that they go hand in hand. There was a time where I felt like I always had to apologise for being organic, apologise for farming with nature and liking trees… But the dialogue has changed, the penny is dropping for a lot more people as well. I'm not so sure it's dropping for every farmer yet, but probably for the ones that perhaps we'd want to surround ourselves with, I think it has dropped.

'But on the other hand,' Geraint says, 'I'm on the board of directors for Natural Resources Wales, and I get the science and the evidence showing how real climate change is and the need for us as a farming business to adapt to the climate. I get challenged a lot about farming methods. There are some good practices going on and some bad practices, but we need everyone on the same page if we're going to address climate change and lead a nature recovery.'

When it comes to income, subsidies make up between 55 and 60 per cent of their income, and then the rest is food production – beef and sheep – but Rachael tells me the critical thing that has kept them viable is having fewer animals but not necessarily less food: 'We realise that we can make more money by having fewer and focusing more on the carcass [quality]. It's having more from less, which I'm a big fan of: being more productive with what you have, being more efficient. I sit in a lot of meetings with NGOs and sometimes government; they think that efficiency and productivity equals more stock. They don't understand it can be more from the same or more from less.'

Yet on parts of the farm, especially the higher ground, it's not simply a question of four hooves bad, more hooves worse. It's which animals and when. When Geraint's parents were farming here the hill ground started to go backwards in quality because they just turned out sheep there.

'The sheep were just there for the summer duration,' says Geraint. 'Now we've got the ability to graze one block all year round and others for different months. The cattle, they go up first of April to the end of November. They are similar to a horse, basically, or a pony. They are pioneer grazers – they will break into areas that weren't being grazed – and then the sheep go in and follow them. When we go up there, you'll see it's 50 per cent lovely, nice and dry, and then the other 50 per cent is recovering blanket bog. The animals are doing much better there and the environment is enhanced for the likes of golden plovers. It's a landscape I absolutely love.'

This chat is the springboard for the farm tour. Rachael and Geraint believe they are simultaneously maintaining food production, increasing wildlife and storing more carbon while steadily increasing profit, but I am keen to see for myself.

Don't let the earlier reference to Snowdonia mislead you – much of their land is lower, softer and more verdant than that mountainous term might suggest. Our first stop is a small triangular field opposite the farm drive bordered by a road, a sharp wooded gully and a disused railway. We enter a strip of a little less than a hectare, where the damp vegetation reaches well above the knee and still-spindly tree trunks reach above head height. It was planted in 2017 with cherry, crab-apple, oak, hawthorn and birch. When the sun comes out, Geraint and Rachael assure me, it's 'moving with bees and insects of all kinds'. Previously it was an awkward corner of the field where the soil was poor, the gradient steep and, in Rachael's words, 'where sheep came to die'. I ask Geraint if there's any sacrifice in food production.

'I don't think we were getting a kilo of red meat off this little corner,' he replies, 'and the time to cultivate it made that just not valuable. Giving it to nature is the best option and it's paid its way with nature, to be honest. When you walk in it, you can feel the soil becoming spongier and healthier. Before it was rock hard, it was like walking on tarmac. It's giving me peace and tranquillity and rapidly becoming my favourite corner on the farm.'

I ask whether his father or grandfather would have come somewhere like this and described it as their 'favourite corner'. Rachael laughs as Geraint replies: 'No, they wouldn't at all. The amount of things that they point out and ask me "Why have you done that? We spent hours taking that hedgerow out and you put it back." They followed a food-producing policy, where they were paid to take hedgerows out and create bigger fields.'

'They'd be horrified,' Rachael adds. 'We've got all sorts of plants like scabious [a wildflower with a bluey-white pin-cushion bloom]. Whereas they thought every square inch of the farm had to be covered in rye grass to be profitable, we see the farm as a jigsaw or patchwork of different habitats and the ruminants work well alongside it. You can have areas that are dedicated to nature that overall help your business resilience as well. We got grants from the Welsh government, which covered 60 per cent of the establishment costs, and then we get £350 per hectare for maintenance.'

A public footpath also runs through this field, so the natural delights on offer are accessible to all, not just the landowners. Rachael and Geraint wanted to see a cycle path along the old railway, linking Bala with a river water sports area about 6.5 km (4 miles) away, but the £1 million quote for the work was considered too steep. Geraint was disappointed: 'Activeness is a must for a healthy nation and there would be nothing better than coming down here in the morning and cycling to school with the children – and there are quite a few children who come from the town to the school just three miles away. Being an ex-railway, it's virtually flat all the way and you can't often say that in Wales.'

Our next field of interest is a short walk along the road and, as the cars dash by on the wet road, Rachael tells me she was 'a thorn in the side of the council' last year. She lobbied them not to cut the verges so harshly all the way back to the fence. They conceded and now just trim the margins so it keeps the sight-lines open for safe motoring but preserves the verge habitat, which today also provides a little food: dessert in the form of wild raspberries.

Our next stop is all about food for animals: pasture. Without it there is no livestock farm. But to understand what Rachael and Geraint are doing on their patch we need to know what's 'normal'. In crop land we know the dominant species are wheat, rapeseed, barley, potatoes, beans, sugar beet and maize, but what are the plants in that green mat being nibbled by sheep and cows? Ryegrass is the big one. It generally grows fast, has high sugar levels and responds well to nitrogen fertiliser input. Decades of breeding – not least from the Aberystwyth Plant Breeding Station, which became the Institute of Grassland and Environmental Research – perfected a range of ryegrass varieties for different conditions. It can be grown long for hay or silage but often becomes a monoculture, which is very poor for wildlife. Other common species mixed in with the sward are timothy and cocksfoot, which have a deeper root structure and so deliver greater drought tolerance. I ask Geraint what's different in the green, green grass of his home.

'Since we took over from our parents,' he replies, 'we've been putting in red clover and white clover as they are nitrogen-fixing and feeding the ground for us instead of needing bought-in inorganic fertilisers. Make it grow well without being dependent on human-made material. The farmyard manure and slurry play their part in feeding [the grass].'

Rachael adds: 'The idea is to be as circular as possible, utilising waste and minimising purchased inputs. It makes sense in terms of profits but also "planet" as we are not reliant on energy-heavy inputs.'

The 14-hectare field that we are standing in is focused on food production, but that isn't to the exclusion of carbon storage and Rachael says they have the numbers to prove it: 'We can't keep saying we're doing the right thing – we have to back it up with data. Since we've been here we've been very rigorous in terms of soil testing and we know we are steadily increasing our organic matter and soil carbon content.'

'We've got a red clover dominant sward here,' Geraint adds, 'and we've taken three cuts of silage in the year [cutting the field when it's tall and green and storing it in big round plastic wrapped bales as cattle-feed] without any inputs, just by closing it off from grazing for six to seven weeks. So it is producing an awful lot more than what it used to.'

This seems to be core to a smarter farming practice. As we walk a little further from the road I see some ruddy-purple clover flowers among the grass turf beneath my feet. So now they have more feed coming from the same field, with no artificial fertiliser input and greater carbon storage. Geraint fleshes it out: 'Now it is established, this field is drawing into the ground about 150 kg of nitrogen per hectare per year. So that is basically free nitrogen from the atmosphere for us, and it grows without having to be fertilised in any way, and it lasts three to four years.'

I ask how much more feed he thinks they will be getting from this field now than previously. 'Three times the amount,' he says. 'One thing I haven't mentioned is that the silage you get from here is at least 19 per cent protein so you don't need to buy in concentrates either [high nutrition feeds]. It's just rocket fuel for an animal. The soil carbon and organic matter have both gone up as well. They were measured quite a few years ago. The carbon was at about 90 tonnes per hectare. Now it's at about 160 tonnes per hectare. And then the organic matter, if I'm correct, is about 14 per cent on this field, which is way beyond the national average.'

In terms of wildlife, the bees love the clover and below ground the worm activity is booming with all that organic matter. I'm so excited by all the different jobs this part of a Welsh valley appears to be doing so I ask Geraint directly if this land is growing more food, storing more carbon and nurturing more wildlife, and his answer is a resounding yes to all three.

Similar ground makes up around 85 hectares of their farm, but they are only hoping to expand this clover mix to cover about 10 per cent of that. There are three main reasons for this: it requires more management than regular pasture as it needs reseeding every three to four years, which adds up to more expense with labour and tractor time; it is such an intensely nutritious feed that it wouldn't be healthy for the sheep to be on here all the time; and it can have side effects for livestock fertility, so they don't graze sheep on it for six weeks prior to conception. But these caveats don't negate the main advantage: their lambs fatten faster on these fields, so they can sell them directly for slaughter and get a better price than if they sold them to be finished on a lowland farm. This also means lower greenhouse gas emissions as the harsh truth is that a shorter life means less time emitting methane and CO_2.

As we leave the field Geraint reveals that there is a downside: they've had to split the field in two with a fence and, in a good winter, this was the favourite sledging slope for the people of Bala. Now there's a bit of jeopardy in whether you'll slide into the wire or be able to chicane through the gate.

We walk through a gate in a hedge line adorned with occasional mature oak and ash trees into the next field. This is shaped like a quarter of a bowl, with the steeper sides curving off to a flatter bottom towards the road. It is a little over 4 hectares acres in total and is about to see a shallow plough as they turn over the top 5–10 cm of this basic pasture and reseed it with eight species of grass. This so-called 'multi-species ley' will be better for food production and biodiversity. They also plan to split the field with a new hedge running along the bottom of the steep part of the bank. The vegetation in the hedge itself and the wider margin will help capture rainwater run-off, thus making the bottom half less soggy, reducing soil erosion and the risk of flooding downstream.

The work will also be addressing a more immediate risk. This whole field is normally left to grow long and cut for silage. When Geraint cut it a few days earlier he had a scare on the steep parts, where the tractor 'wanted to go over'. Farming is a notoriously dangerous profession. According to the UK Health and Safety Executive (HSE), agriculture has the worst rate of fatal injuries of all the industrial sectors, and overall injury rates are twenty-one times higher than the average for all industries. (Given the rigour of accident reporting on most manufacturing or building sites I have visited vs the more casual attitude on family farms, I wouldn't be surprised if the real figure is even greater.) Tractor rollovers are a big risk, with around 130 such incidents every year in the USA. Slopes are especially hazardous and, when combined with the G-forces of turning sharply a little too fast, vehicles can soon turn over. In the year 2022–23 the HSE recorded the death of a farming contractor in Wales who was lime-spreading on steep ground.

'If somebody else was helping me out here and they had an accident, I couldn't forgive myself,' says Geraint. 'So we'll take mechanical use of this top part of the field out of the equation totally and just graze it. We can grow forage [and use machinery] in the flatter bottom part where it is safer.'

This approach is remarkable in a number of ways. Firstly, realising that farm safety isn't just about being careful but eliminating some of the system risks in the way that you till the land. Then there is the act of planting a new hedge, which goes against the grain of late twentieth-century farming. So many have been ripped out, and even now it is quite rare for a new one to be introduced. Here it is being done for productive farming reasons, not simply for the birds and the bees. Overall, they have put in 14 km of hedgerows, some of that subsidised by payments for 'ecosystem services' from the beer company Stella Artois brokered by the National Park. (Elsewhere the National Grid pays for the restoration of stone walls and the creation of fences in compensation for the ugliness of existing pylons.) But what's most impressive is the extraordinary attention to detail Geraint and Rachael pay to every part of their farm – management on a granular level to get the most out of every square metre. This is intensive farming how it should be: the intense application of intellect. As Rachael says, 'You need a holistic total farm view, but you've also got to know what works best for each part of the farm. It's about optimising production where you can, but managing other parts for trees or nature or lighter grazing. We can't have a blanket approach.'

Geraint adds: 'I've had five soil-nutrient plans on the farm over the last twelve years, identifying nutrient levels in every single field and seeing how they've changed. It has to be managed field by field so we can get as much out of them whether it be for food production or the environment.'

This is in marked contrast, they suspect, to the majority of hill farmers – not least his parents, who left the farm in 'a sorry state' according to Rachael with soil, boundaries and infrastructure lacking investment. Geraint agrees.

'You've got a lot of farmers still dreaming of the good old days when they were paid to produce food, plough fields, plough hillsides, and they just long to get back to it,' he says, reflecting on the single-minded approach of many farmers. 'I remember coming out of a meeting of the Farmers' Union of Wales and they knew I was going down to London the next day to meet Michael Gove [then Secretary of State for Environment Food and Rural Affairs]. Someone grabbed my arm and said, "You tell that Gove to forget about the environment. We are food producers and that's all." And I said, "What do you mean? That is not what we should be doing as farmers. We should be approaching it as an entirety, where we are delivering multiple things from a farm business." With modern technologies, we can produce as much of the food as we need here in the UK and care for nature, but we just need to adapt our approach.'

Rachael also firmly rejects the idea that it has to be either/or – either food or the environment – but she says the balance between the two will shift in different parts of their holding and the country: 'We have 80 hectares that is mainly rock and bracken, and we think probably there the animals will be almost a by-product and we'll probably retire there and live in a really lovely place. Overall, it's about understanding where we can optimise food production and where we can optimise nature. And there'll be some farms like this one, where we can do both, and other farms where one will be much stronger than the other.'

As we've been talking, the rain has got harder but we've still got the higher hill to see. It's only accessible by quad bike so we return to the farmyard, where Geraint and Rachael rapidly load the ATV onto a trailer with the speed and efficiency of a Formula One pit stop. Before we leave, though, there's time for a walk in the woods past a couple of ponds they've dug to provide water for the cattle when they are housed over winter. They are now lined with bullrushes and provide habitat to geese, nesting ducks and a heronry. But in recent years it's been running dry regularly even though it has a sizeable catchment. They both suggest this could be climate change at work.

The trees are mostly mature oaks leaving grass beneath and plenty of rotting wood piles. Geraint and Rachael use a fraction of what is available for firewood, while the rest is stacked to decay as Geraint says decomposing timber is as good for wildlife as the growing stuff. Rachael observes that, despite its extraordinary natural value for wildlife and as a carbon store, it's not rewarded in current government policy that claims to support the environment: 'On paper, this is worthless financially. We are tree-huggers. We think it is great. But for cash, you'd cut all these trees down and plough it and produce food, which would be more financially viable. So we need to start shifting [policy] values to this sort of habitat.'

On the top side of the copse, about 300 metres above sea level, is a rarity at this altitude: cropland. There are 11 hectares of future animal feed here and they've got a varied trough-full coming their way: peas, oats, barley and multiple grass species. Another 'rocket fuel' forage, according to Geraint, which avoids the need for buying in concentrated feed: 'Growing crops this high up is unusual, is bucking the trend. But we used to do it [historically] and we've forgotten that we used to do it.' According to Rachael, that past crop was the biofuel of its day.

'Most farms like this would have grown oats years ago for horses and cattle,' she says. 'You can see arable field names on the old tithe maps. But now we've got into this ryegrass trend, thinking we can only grow grass. But the peas in here make it a high-protein ration, which reduces costs and finishing times for stock. We alternate between growing swedes for feed one year and this stuff the next. But with such volatile weather, it's an annual experiment.'

Returning to the yard, we jump into the pick-up towing the trailer with the quad bike. The rain gets heavier as we snake our way out of the valley onto the open moorland. Trees fade away and the views open up. But, aside from dramatic geomorphology, there is not much going on here. I ask Geraint if he thinks the land is degraded and he chooses his words carefully, emphasising the future potential rather than the current state: 'It's just ticking along, delivering a little bit here and there, and this approach is very common, which is a shame. But what it can do food production-wise and caring for nature-wise... there is so much more you can do here. The true ability of this place would be fantastic.'

Impressively, Geraint is delivering these big thoughts while reversing a trailer into a tight spot between two crumbling stone walls. We board the quad bike and traverse a little more than a kilometre (around two-thirds of a mile) across a gently rising plateau of heather, rough grass and occasional drainage ditches scarring the peat. This is not their land. To the untrained eye, 'featureless' would be unkind but not entirely untrue. Then we reach a boundary fence and open a gate into the part they own. It's not like suddenly stepping into a jungle – this is still a grazed land-scape – but it immediately feels more abundant, with thicker and more varied ground cover.

In his parents' time 1500 sheep were 'pushed' up here onto 260 hectares. It was overgrazed. His and Rachael's management strategy has been to block up some of those gashes in the peat to prevent the ground from drying and emitting more CO_2, to have fewer sheep but with better genetics so they get more red meat from fewer animals, and to increase the number of cattle as the way they graze enables greater plant variety. The cows are mainly Welsh blacks, now just visible as our eyeline breaks a ridge into the next gentle fold. But catching my eye immediately at my feet is a superfood ready to eat: bilberries. They are squat hardy bushes with small waxy leaves and dark berries, their sweetness betraying the sunshine of early summer, not the murk of today. Geraint says the way the cattle graze opens

up more space for the berries before they get a late-summer
visit from the sheep: 'Come September, because it's the end
of the growing season, the plant is putting all its energy into
its leaves so the protein levels are higher and they really like
nibbling on it. So it's quite good to have a handful of lambs
from here that we put in the freezer ourselves. The taste is
ten times better.'

Resident wildlife is largely invisible today, sheltering
from the persistent rain, but Rachael and Geraint assure me
it's humming with bees in the sunshine, home to ground-
nesting birds and the territory of a short-eared owl hunting
mice and meadow pipits among the scrubby vegetation.
Also increasingly appearing among the thicker ground
cover are small trees: mountain ash and self-seeded conifers
from the nearby timber plantations. Once grown they are a
mixed blessing as they provide an inviting perch point for
raptors seeking cherished golden plover chicks below – a
predator's watchtower.

Here again, land uses are stacked: food, carbon
storage and wildlife. But that, according to Rachael,
is not straightforward to achieve: 'We are keen to break
down the misconception that it's sort of "dog and stick"
[unsophisticated] type farming. That it's one-dimensional,
that it's quite simplistic. It's a landscape where we can stack
enterprises and deliver many functions.'

As I look across the landscape, my untrained eye can't
distinguish with certainty how the bulk of this land is being
managed. Geraint says that, while there are pockets of
farmers working like him, most are still more traditional hill
farmers. Both he and Rachael cherish the variety of what

land can deliver. They are proud of the taste and nutritional quality of their red meat, they are thrilled by the results of the latest bird survey revealing between seventy-five and eighty different species, and they rest easy in the knowledge that their land is being improved not degraded: 'We've got a lot of species that depend on co-existence with grazing animals, and the best thing about it is that food is being produced here as well.'

I've learned a lot, but I am wet to my underpants. Time for a shower, a change of clothes and a welcome sandwich before hitting the road with plenty to occupy my mind.

For these two farming chapters, I have parachuted in on a small handful of brilliant farmers, but there are professionals out there who have worked with hundreds of land managers: agronomists and farming advisors. Niels Corfield is one of the most respected experts on regenerative farming in the UK. He does 'on farm' monitoring work on over a hundred properties and is convinced that the kind of farming practised by the farmers we've met in these two chapters, which builds the natural assets of the land, delivers more of everything: food, wildlife, water and carbon storage. Fertilisers, pesticides and herbicides are not totally ruled out, as he is more interested in outcomes than ideology, but he firmly believes in putting all these chemical inputs 'in the crosshairs' and asking if they are justified: 'It all boils down to maximising gains in natural capital, because yield springs from the well-spring that is functioning ecosystems – healthy soil, water, air and other free resources.'

I put it to him that farmers in recent decades have shown little interest in soil life, little interest in biology. He generally agrees, albeit with some ticking-off for my journalistic penchant for generalisation.

'I try to avoid the truisms and memes of "All farmers do, want or say such and such". But we can say categorically that the foundational trials done in the early to mid-twentieth century that established fertiliser response in crops were done in inert, sand or sterilised soils to "remove variables". But the variable they removed was the entirety of biological function. So of course you are going to see excellent growth response to inputs, as the plants had no other way to access nutrients. You've removed their natural partners: diverse organisms that live on or in the roots. This led to a false conclusion: as you get pretty guaranteed growth increases from applying nitrogen, then this is the only way – the sole means – to get that extra yield.'

Looked at from the twenty-first century, with our apparent sensitivity to natural relationships, this seems staggering to me. The soil was being treated as little more than the holder for the plant, not its integral partner – a mere footing, not a vibrant home. It's like an alien encountering a human being and thinking, *How do we grow that? All that diet and digestion stuff looks a bit impenetrable. Let's just take a protein, sugar and vitamin mix and feed it intravenously.* The result would be similar to the human batteries in the prescient film *The Matrix*, where sentient machines store banks of somnolent people to provide electrical power: it works for sustaining a life form but not a life.

I say 'apparent sensitivity' on the previous page because maybe our current respect for soil ecology is merely lip service uttered by the public relations side of farming unions and chemical companies and then repeated by the likes of me. Is the whole edifice of commercial farming really built on these narrow-minded experiments? Before I hear more from Niels, I want a second opinion on this so I speak to Anna Krzywoszynska. She is an author, anthropologist and soil scientist, meaning she looks not only at the chemistry of the ground beneath our crops but also the human decisions underlying it.

'The way that we understand soils in conventional agriculture,' Anna says, 'starts with the German chemist Justus von Liebig, who first proposed that for crop growth all you have to contend with are the limiting factors for crop growth. It doesn't matter so much about what is in the soil so much as what there is not enough of. And you add more of these missing elements from chemicals rather than organic matter or humus. We are in that mindset today.'

Anna suggests that chemical dominance increased further due to specialisation in farming, particularly the separation of animals and crops. When most farms were mixed, farmers had to handle manure and knew it was a biological product but, once the livestock vanished, fertility support was simply chemical. Given the recent reverence for soil in farming, where does this leave the science?

'Soil biology itself is only a recent concept,' she says, 'it comes in the bracket of "un-done" science. We just weren't interested because chemical farming had all the answers. Organic farming has been seen as "unscientific" because it has not been of interest to mainstream science. This is not the fault of organic farming itself.'

There is an emerging industry producing biological products to replace chemical ones, but Anna believes it is still stuck in the same 'additive mindset'. There is a lot of research into what are known as 'bio-stimulants': 'Bugs in jars replacing chemicals in a sack. But still enabling big companies to shift products. This is not changing the system.' She would like to see much more groundwork: 'We should be studying the soil on a particular field and working out which crop types are best suited to a growing partnership with the biological inhabitants of the soil. But it takes time and you are not necessarily selling anything at the end of it.'

There is no doubt that the companies selling the chemical products have enormous influence in farming. Chemically reliant farming has emerged as the dominant wisdom because many of those dispensing that 'wisdom' are agronomists – experts in the science and technology of food production often engaged by farmers to keep their businesses in shape. Many of them receive more money if they sell more chemicals, just like some doctors who get a bonus for prescribing pharmaceutical drugs, which leads to a similar suspicion: their diagnosis of what the patient or the farm requires is being strongly influenced by what makes them the most money. Yet nearly all regenerative farmers want to use much less fertiliser and biocides, while not necessarily abstaining totally as with organic production. So it is in many agronomists' interests to be gatekeepers of conventional agriculture while trying to stop regenerative practices getting in. Tim Parton, the regenerative arable farmer we heard from in the previous chapter, split from traditional agronomy but knows how painful that break-up can be.

'When I give talks,' Tim tells me, 'some people come up to me and say: "I love what you are doing but my agronomist says it won't work for us." I reply: "change your agronomist," and they reply: "I can't do that, he's a family friend." Back before the seventies, agronomists didn't exist and now they've done so well that the farmer isn't the farmer any more. It's the agronomist who tells them what to grow, when to plant it, what to put on it, when to harvest it. Who's the farmer? They have lost the ability to farm and make decisions for themselves, and the chemical service agronomist has taken that role. And that's why they get invited to family weddings, funerals and so on. They become a member of the team. But the minute they stop paying them, they are not a friend at all. People phone me up and say: "You were bloody right, they haven't been near since. I thought he was a friend."'

I ask if it is like being friends with your local drug dealer.

'Yeah, basically,' Tim replies, 'because that person has got to sell. They are working for a big agrochemical company. That company has bought 70,000 litres of chemical. So it's got to be sold. So it will be sold. It's very easy to convince a farmer when you open the magazines and you read "It's going to be the worst *Septoria* year ever" to sell him that chemical to guard against that risk.'

When Tim alleges that companies who sell chemicals pay to put scare stories in the farming press, I express some doubt but he says: 'They all pay to put content in the agricultural papers and magazines and it does not say that it is paid for.'

Niels Corfield thinks there is an associated problem alongside the financial motive: their very limited toolbox, which is just a chemistry set: 'If all you have is a hammer, everything looks like a nail. All they know to do is to prescribe nitrogen and fungicides, for example.'

Niels, who is himself a farming advisor, believes there is another damaging fallacy about soil nutrition: the idea of 'soil mining', which suggests that – unless you are applying chemical additives from the outside – you are steadily diminishing what is there.

'The notion is [based on] the law of return,' he says. 'The idea is that as you are taking away [nutrients] through cropping or grazing, you need to put back. A sort of "one in, one out" approach, which is all-pervading in conventional agronomy and ignores biological and mineral factors. But we regenerative advocates think that soil biology – mainly fungi and bacteria – has incredible ability to liberate and solubilise those nutrients and transfer them from the soil to the plant. They can dissolve sand, silt and clay through digestive processes – enzymes and strong acids – the same way that we digest food. They become assimilated into their bodies and, from that point onwards, they are available to plants. But the problem for conventional arable operations is that we have broken the relationship with soil biology to the point where the useful organisms are so depleted that, when you remove the [chemical] inputs, you see massive declines in yield. In essence, the regenerative approach is trying to fast-track the restoration of the natural nutrient-cycling process. It's about building up microbial populations, restoring the relationship between plants and organisms, and removing practices that sever that link. But it isn't happening overnight.'

Indeed not. Depending on soil type, its level of biological bankruptcy and the expertise of the farmer, it can take years to transform a conventional farm into a thriving regenerative holding. In the meantime, income may slump, your familiar agronomist is tutting, your senior family members are questioning your sanity, and challenging emails from the bank are piling up in your inbox. It's widely agreed that coming off chemicals is painful for profit and yield.

However, that isn't the end of the story. Niels has worked with seasoned regenerative farmers in the UK and talked with many of the godfathers of the practice in the US and Australia. Through observation or conversation he has learned a similar story: once the symbiosis between bugs and plants is restored, the reawakened and multiplying soil life gets to work on naturally occurring minerals. The farmer, quite literally, reaps the reward.

'What they see is steady increases across the board in available nutrients. So rather than mining soil that is the standard criticism offered by orthodox agronomy, what we see is exactly the opposite. Every hectare of randomly selected fields might have 1.5 tonnes of phosphorus, 10 tonnes of nitrogen, 7 tonnes of potassium in just the top 10 centimetres of soil. So what we need to do is partner with those organisms, bacteria and fungi that have the ability to unlock those reserves and make them available to plants.'

Regenerative farming is not just for arable farmers; many graziers have driven the movement and many practitioners champion the traditional mixed farm with a return of animals to arable lands. Niels thinks pastureland has huge potential to be regenerated because so much of it is in a bad state due to overgrazing, soil compaction and monoculture grasses.

'I'd say high 90 per cent of all UK pastures are "bonsaid" like those tiny trees. The same genetics and seed varieties could be four times the size in terms of root mass, the crown and leaf size. We could go from these one-inch-high specimens into six-inch tall, twelve-inch roots, deep and dense and tall and fat. With that comes excellent soil structure, excellent infiltration rates [the speed that soil can absorb surface water], leading to greater water-use efficiency.'

That may sound a bit abstract but the water-holding ability of the ground is one of its most fundamental qualities. Niels is so keen to drill home its importance, he reaches for a truism himself: 'One of the few universal truths in agriculture is that the irrigated crop outperforms the non-irrigated crop. So more water stored in our soils will translate into more yield. We also remove flooding issues, we remove erosion issues and we remove drought issues as the water that was just running off is now in the land. It can then be drawn upon by the plants in hot dry periods leading to additional production.'

He believes that regenerative farming practices with deep-rooted grasses can turn what is seen as one of the greatest threats of a warming climate – intense heatwaves – into an asset: 'Droughts are our biggest opportunity, because high temperatures and long sunlit days with no clouds will grow a crop better than anything out of a can.'

Overall, Niels has great faith that regenerative farming can help us to win the space race. It's small now as a proportion of the world's farmed area but it is growing fast and, in that area, he has seen proof of multiple wins: 'As you regenerate soils, you regenerate ecosystems because it is ecosystem processes that are driving the restoration of soils, and with it comes an increase to the base of the food chain for insects and therefore birds. So we should see uplift on all metrics if it is being done right.'

Yet there is part of my argument that Niels disputes strongly – he doesn't think we need to grow more food: 'I wholeheartedly disagree that there is a food emergency or a production issue, and the correlation between food production and food access is weak at best. Food access is a social and economic issue, it is not a production issue.'

This takes us into the arena of human choice, and Niels is right insofar as smart land use isn't simply a calculation of population, calories, tonnes of carbon and natural capital metrics. It is intricately linked with diet, wastage, equality and distribution. These behavioural changes could have huge land use impacts, as we'll see in chapter 10.

6

PEAT AND CARBON

The farmers and cheerleaders for regenerative farming we heard from in the previous chapter focused on the dilemma of whether we can produce as much, if not more, food from the same area of land while reducing the use of fertiliser and increasing the capacity of that space to host wildlife. But there is one farming location that really gets to the heart of the argument on food vs climate. It's not actually one place on the planet but one type of soil, common across much of the most productive land on the planet: peat.

Getting passionate about this variety of deep black dirt is a particular trait of farmers, gardeners and environmentalists. It's either very good to use or very bad to use, depending on your perspective. Peat is the accumulated remnants of partly decomposed plant matter built up over millennia in slightly acidic fresh water. In the uplands of Britain and much of Europe the biological building block is sphagnum moss: a greeny-red spongey plant at home in bogs. In the lowlands, like the Fens of eastern England or the Dutch polders, peat is what's stored beneath as the marshland grows above. As it is almost entirely composed of ex-plants, it has a very high carbon content and can be considered a halfway house between living plants and fossil fuels like coal or oil. The carbon in peat is locked away by something both simple and vulnerable: water. All that's needed to allow the carbon to slip the leash and become global-warming CO_2 is drying the peat by draining the marsh. And that is something humankind has really warmed to; we began hundreds of years ago and have accelerated since. The Fens are a low-lying area of eastern England with much in common with the Low Countries across the North Sea, as ten thousand years ago a land bridge joined the two and the rivers of this part of England drained into the Rhine. In the seventeenth century, a flow of people came the other way, as Dutch drainage experts helped rid the Fens of that pesky water.

The motive was simple then and remains so today: peatlands make great farmland. Half of all the grade-one farming land in England is found in the East Anglian Fens, though they occupy just 3 per cent of the country's land area. This peaty production hub delivers just under a quarter of crops and just over a third of the country's vegetables. So it's a breadbasket and climate burner in one.

John Shropshire is executive chairman of G's Fresh, a huge salad- and vegetable-growing operation with farms in five countries (the UK, Poland, Spain, the Czech Republic and Senegal), supplying retailers on both sides of the Atlantic. Its name derives from his father, Guy, who began the enterprise from a Fenland farm near the ancient but small city of Ely in Cambridgeshire. G's East Anglian holding now spreads across 5,000 hectares in the Fens, some owned by others but all under their management.

The productivity of peat was the foundation of John's family business, but what was once a blessing is beginning to feel like a curse. John cares about climate change – he can see its impact on his farming operations and knows those same operations are driving it. He tells me: 'Farming the thinner peatland soils can emit 8 tonnes of CO_2 per hectare per year, whereas the deep peat could be 30 tonnes per hectare. Farming in the Fens is likened to cutting down rainforest because of peat loss. I am spending half my time worrying about that.'

By 'loss' he means mainly the oxidation of carbon in the peat, which creates CO_2 and causes the disappearance of the ground beneath his feet. In some areas this has been worsened by the soil itself simply being blown or washed away. In the 1850s a group of landowners wanted to measure this decline and so they drove a long post through the peat to sit on the base layer of clay. Today these 'Holme posts' reveal the land has dropped by four metres. That's equivalent to two storeys of a house – across thousands of square kilometres – that have just vaporised and are now helping to make a thicker blanket round the world. Visualising the absence of something can often be tricky, but this particular absence is also revealed as you drive around the Fens: so often you are looking down on the fields as the road level has been artificially maintained while the land around it shrinks. On minor routes, the road is also subsiding and you are left with a driving experience like mogul skiing: bouncing over the rolling asphalt. Navigating these roads has some novelty but it can be lethal as cars begin to fly and the accident statistics mount.

John has witnessed the lowering of his own land: 'In the forty years I've been farming it's dropped by at least half a metre. This is mostly down to oxidation of the peat as we successfully keep wind erosion to a minimum by planting cover crops and rarely leaving bare fields.'

The vanishing of the very ground we need to farm triggers dire warnings that there are fewer than fifty harvests left before the black gold has gone. The number isn't precise but the message is clear: current farming practice in our breadbasket is not sustainable.

And then you have the climatic harm. The average British citizen is responsible for 8 tonnes of CO_2 per year – an equivalent amount of global warming to the figure John gives for farming a hectare of the thinner peatland soils. In Europe, peatland farming emits a quarter of agriculture's greenhouse gases but only occupies about 3 per cent of its farmland. Globally, our use and abuse of peat contributes about 5 per cent of greenhouse gases. However you score it, it's a big chunk of our climate change problem.

John is open about the impacts and importance of farming. 'Being a very big farming operation, we have a massive impact on the environment worldwide and communities too,' he tells me. 'We are farming, we *have* to impact the environment. You have to destroy something that was there before. What we try to do is *lessen* that impact. Our aim is continuous improvement.' To that end, G's Fresh is a partner and funder of the Fenland Soil, a not-for-profit grouping of farmers and academics who aim to inform and develop 'whole farm' land-use policies to limit climate change and boost biodiversity. The maintenance of some farming is implicit. They have funded four flux towers – platforms that sit above fields and are adorned with kit to measure the exchange of gases in and out of the farmland beneath. But farming practice is changing too. Some of the deeper peats could be taken out of production and flooded once more – rewetted, in the jargon – so that the return of the water imprisons the carbon again. Where most of the peat has gone and the remaining layer is thinner, they will use some of the regenerative farming techniques detailed in the previous chapter: reduced ploughing, cover crops and strip tilling. But John reckons that this will reduce crop output by about 10 per cent and, while he's agreed to this, he's also troubled by it. It's a stone in his shoe.

'The best soil should be used fairly intensively to produce the most food,' he says. 'That is a responsible use of land. And, although I know that farming the deep peat releases a lot of carbon, we use no nitrogen fertiliser on it and I *would* have to if I moved production to other less naturally fertile ground. No one has calculated the climate impact of that. We think you can farm the Fens sustainably. This won't eliminate but steeply lessen impact.'

He also questions some of the revered guidance on climate matters. The Climate Change Committee's most recent Carbon Budget recommends 'rewetting and sustainable management of 60% of lowland peat by 2050.' But, John argues, 'that doesn't include the carbon impact of growing and importing the lost food production. Government scientists say it will be made up by vertical farming. But that is incredible, implausible, as the biggest vertical farm in the world, in the US, worth a billion dollars, last year produced

800 tonnes of salad. Yesterday we produced 800 tonnes of salad across our farming estate *in one day*! And on top of that, I'm not sure anyone has calculated the amount of electricity to produce a carbohydrate [indoors]'.

This relates to the issue about whether or when it is desirable to reduce food production locally and import it instead. But John knows a thing or two about the economic and environment cost of food imports as he is responsible for truckloads – even shiploads – of it.

'There are a couple of certainties: we need more research, and distance in itself is not as big a deal as people think. The carbon footprint of radishes and spring onions from our farms in Senegal is low as the soil needs little fertiliser; water* and sunshine are plentiful; and 98 per cent of what we grow is transported by ship. In Spain, where we grow salad, we need no artificial warmth but some of the water comes from desalination plants. These are massive energy users but the resulting CO_2 depends on the proportion of renewables on the grid. In Spain and Portugal this is increasing. The produce comes from there by conventional diesel-powered truck. A round trip of over 4,000 kilometres. But many of the salad vegetables grown in the UK, especially away from the summer months, require heated greenhouses, usually with fossil fuels and this is the big one.'

The respected resource and website Our World in Data backs up John's argument that what you eat is more

* However, a report from the World Bank describes the country's water resources as 'deteriorating and inadequate' and specifically on farming, 'surface water is the main water source for agriculture, but during periods of low rainfall, there is not enough to meet demand, especially for irrigation water in the Senegal River Basin'.

important than where it comes from. They even say that 'Eat Local' is one of 'the most misguided pieces of advice'. They quote research suggesting that for American households the transport of food contributed only 5 per cent of its total climate impact – or, viewed another way, if they all lived right next to the farm where the food was grown it would only reduce their dietary emissions by a maximum of 5 per cent.

They also pull together studies that specifically support the assertion that eating locally can increase emissions, saying it is estimated that: 'Importing Spanish lettuce to the UK during winter months results in three to eight times lower emissions than producing it locally. The same applies for other foods: tomatoes produced in greenhouses in Sweden used 10 times as much energy as importing tomatoes from southern Europe where they were in-season.'

This is, of course, an average and some growers in northern Europe are trying to heat their greenhouses with lower-carbon sources. The big transport polluter per kilo is air freight – around fifty times greater than shipping on a ship – but it is rarer than you might think, accounting for just 0.16 per cent of food miles. If you want to steer clear of air-freighted foods (a pretty small sacrifice for a big climate gain), watch out for highly perishable fruit and vegetables like berries, beans or asparagus. I'd like to see a mandatory air-freighted label. You should also be aware of growing seasons. According to Sarah Bridle in her book *Food and Climate Change Without the Hot Air*, a dish of homegrown French beans in season has seven to eight times less climate impact than a plateful flown in out of season.

Yet, there is another factor to consider in imported food (besides energy to grow and fuel to transport) that is especially relevant to space-hungry staple crops, as John points out: 'With crops like wheat, maize or soy, there is a real risk of deforestation.' John believes that if he grows less of these commodities, the UK puts more demand for them into the world market, which then incentivises more farming on formerly wild lands, thereby reducing space for nature.

In April 2023, I go along to a two-day conference dedicated to peat and farming. It's held in Ely, the small Fenland city mentioned above, also known as 'The Isle of Ely' as it stands on slightly higher ground and, before the surrounding wetlands were drained, was an island. The massive cathedral still dominates the skyline today, and it offers an iconic approach to the city when travelling by road: with the land so flat and the planners doing their job of keeping the other buildings low, God's big house can be seen from pretty much everywhere. It's an apt location to debate farming's impact on the heavens.

The first thing that strikes me is the size and variety of the congregation for this conference: about 160 people – 'two years ago we'd have been lucky to fill the front row,' comments one speaker – comprising academics, civil servants, local politicians and farmers. Lots of farmers, which I can tell just by scanning the crowd as so many are sporting Schöffel waistcoats, fleece gilets with leather trim round the zip and arm holes: agricultural plumage. This is impressive; farmers are busy people – there are always jobs to do on the farm, especially on a sunny spring day – and they've left their land to be told bad news: 'business as usual' won't do. But this group have most definitely not got their heads buried in the sand or the vanishing peat. They can sense the threats both short and long term and the speakers spell it out for them.

The large attendance is not lost on Chris Evans, a biogeochemist from the UK Centre for Ecology and Hydrology, who tells me: 'It makes a refreshing change. When we first started this work a few years back, many of the farmers were so suspicious we had trouble even getting on their fields. They felt that they weren't in the room for a conversation that was being held between scientists, lobby groups and government. All they heard was headlines saying "We've got to flood all the peat and do it tomorrow". At the same time, on the ground, we had a few "boutique" restoration projects while the vast bulk of the farmed landscape was just carrying on as normal. That wasn't really tipping the scales on carbon or wildlife.'

Chris and his colleagues made it very clear that they were not trying to end farming on the Fens but wanted to work with land managers to measure exactly what was happening and to develop solutions. At the same time, many farmers had worked long enough to know their soil was disappearing. His presentation opens up with some killer stats:

- Soils hold almost twice as much carbon as the atmosphere and living plants and animals combined.
- Just 30 cm depth of peat holds as much carbon across a hectare as a hectare of rainforest.
- Deep peat can easily hold ten times as much carbon as woodland.

- In the UK, carbon emissions from peat cancelled out absorption by our forests. Indeed, even some of our depleted peatland soils still hold enough to make other habitats look like deserts.

So, given all these inconvenient numerical truths, what is his solution? 'A mosaic,' he says, 'made up of four main objectives: land focused on food production, land focused on reducing greenhouse gases, land focused on wildlife, and land focused on water storage.'

The key to continued farming on peat is to raise the water table beneath the crop as high as possible. The more peat is submerged, the less is releasing CO_2. This is what they are trying at G's, but there are plenty of challenges. Most food crops like their roots to be moist but not waterlogged, as they need a combination of water and air around their filaments to work best. The water table needs to be as high as possible for carbon storage but below expected root growth, which is not as easy as balancing the taps and the plug on your bath; neighbouring farmers and settlements have a say and extremes of drought or rainfall can ruin a best-laid plan. Machinery prefers firm ground and can get bogged down. Then there is the heritage of the crops themselves. Some, like celery, are a wetland plant but, according to Chris, most of our cereals came from near deserts: 'Wheat started in ancient Mesopotamia and then evolved and spread round the world. But it is still fundamentally a dryland crop so you probably have to keep the water table at least 50 cm below the surface.'

There is one farming decision that Chris believes is so urgent and obvious that he sheds his cautious tone: 'I would ban maize production for energy production in anaerobic digesters because it is insane. It's not food, or even animal feed, it is subsidy driven and, in climate terms, it could be worse than burning coal.' Surveys suggest that 4 per cent of lowland peat is used to grow maize and still more for sugar beet, much of which is also fed into anaerobic digester plants. As already discussed, bioenergy crops are generally a very inefficient use of land and this is surely massively amplified when their very growth causes plumes of CO_2 from peat.

At the opposite end of the spectrum is the climate farm, one where the primary objective is to use land to cut global warming by absorbing and storing carbon. The farmer I'm en route to visit calls this 'reverse coal' and reckons he can change his land from emitting 25 tonnes of carbon per hectare to sequestering 30 tonnes per hectare while still producing food, which sounds like alchemy. It is the Lapwing Estate on the Humber Head Levels just to the east of Doncaster. The Humber Head Levels are one of the four key lowland peat areas in the UK (alongside Manchester Moss, the Somerset Levels and East Anglian Fens). Given this low altitude, I'm somewhat confused by the address – Gringley on the Hill – but height is relative and the nearby village was wisely built on the slight rise at the edge of the Fen. My car suspension reveals that I'm in peatland: the land around me is as flat as a lake but, as I experienced near Ely, the road is a rollercoaster. Its peat-based foundation is evaporating and the tarmac is subsiding.

The farm is just a few hundred metres from the Fen edge and it looks nothing special, just open-sided barns and a utilitarian office block surrounded by gravelly hard standing. But I soon discover that this is home to one of the most innovative and original farmers I have ever encountered. Never mind 'the smartest guy in the room'; James Brown is the smartest guy in the field – my description, not his, though he does believe he is in the vanguard of a new trend.

'My current business model,' he says, 'is I grow a crop on the land and sell it. My future business model is I wet the land, grow a woody crop on the land, which becomes a fuel for my vertical farm and raw material for my carbon storage. So the carbon sequestration would be a carbon credit. I would also then be able to enhance biodiversity as a biodiversity credit. I'll be able to prevent flooding [downstream] by allowing this to be flooded. I will also improve water quality.'

That's quite a list, so how will he get there? Currently the Lapwing Estate is a 2000-hectare organic farm, growing fresh vegetables like broccoli, cauliflowers, peas and some arable crops. There is also a 600-head organic dairy business, run on what is called the New Zealand model, meaning that cows are out all year, grass-fed, and the yield fluctuates according to pasture growth. But looking further back lends further plausibility to this great leap forward: his family are no strangers to bold change. They bought this farm in 1994, at which time they owned a very large agrochemical distribution business, selling just the kind of products like fertilisers and pesticides that are specifically outlawed on an organic holding. In 1997 their company had a record year but, rather than resting on their earnings,

James and his dad began to look at potential threats to the business. Successful farming without chemicals is the obvious danger and so, in an effort to 'know the enemy', they decided to see if it was possible. They started out with a small plot but the 'infection' took root and by 2007, when James took over, they sold the agrochemical business and went completely organic.

'Generational shift is always a good time to make big changes. I knew there was some risk but I believed in more sustainable food production,' says James, who grew up on the farm and was always curious about how it worked, although he tells me his studies and early career were a long way from the Fens: 'I didn't go to agricultural college; my first degree was politics and economics. The second was in international relations, and I studied accountancy in London with a firm and then have come back here. So I've come into the farm with absolutely no knowledge of farming at all, which has probably been quite a benefit for this sort of thought process.'

We talk briefly in his office but I am itching to get the guided tour and see for myself. His initial reluctance is explained by the fact that his chunky four-wheel drive is in the garage for a service and the 'loaner' is a somewhat low-slung hatchback that beeps frantically when close to vegetation. Too bad – we are soon ploughing through long grass.

Setting off out of the yard, I comment on eye-catching poppies among a crop of winter wheat. 'Weed control is one of the downsides of organic farming,' he replies, 'but as long as you don't have too many, you won't suffer much of a yield penalty. You can probably guess why we called our daughter "Poppy" who was born in July.'

James was happy with the organic farm, which made a good income, enabled a wildlife recovery and, by changing his tilling, irrigation and food-storage techniques, reduced operational carbon emissions by half. On the downside, his farm was disappearing. I mentioned this vanishing soil phenomenon earlier but it bears reinforcing because it's just about the clearest visual evidence of carbon loss and emissions. James has some potent examples.

'See this bridge?' he says as we cross a brook about a metre across between two fields. 'We've had to re-ramp either side twice in the last twenty-five years. The structure has stayed in the same place but the land either side has dropped. You can measure the drop at 30 centimetres in a generation.'

He warms to his theme as we drive on past a drunken row of telegraph poles. Look down the line and they are staggering this way and that as the land shifts beneath their feet.

'I'm just going to show you a grain store here, which was put up in the 1970s. It was "piled" to take the weight of the grain and put in flat to the ground, to the concrete yard around. But you can see now it's on its own little raised platform. There are ramps up to it and I'd say the land around has dropped close to a metre in the last fifty years.'

This chimes with the research from scientists like Chris Evans from the Centre of Ecology and Hydrology, who estimate that each drained hectare of deep peat is emitting 25–30 tonnes of CO_2 and lowering by roughly a centimetre per year. James knows that the only way to stop this is to rewet the peat. They did this for a few decades on 20 hectares next to the River Idle and it is now a thriving wetland, home to many wading birds, and acts as a water store in times of flood to protect nearby settlements. But James has one serious reservation about this.

'If I raise the water table and it'll stop food production,' he explains, 'and that food production is going to be produced in the Netherlands or in Belgium on exactly the same type of peatland that I've rewetted, what have we achieved apart from adding transport cost? The answer is nothing. So peatland restoration, which does not include food production within your envelope, is… I wouldn't say dishonest, but it is not the whole story. Therefore, unless you model where the output that is currently being produced from that land will have to come from (and include that in your calculation), you haven't helped. You haven't actually saved any carbon at all.'

As we drive through the weeds along the top of the raised riverbanks (the water is higher than the fields), James points out the first stages of his answer to this puzzle: 'Looking down to your right-hand side, there are four 5-hectare fields where we've grown a whole different set of varieties. We've got willow, we've got miscanthus [a grass of 2 metres plus], we've got alder, we've got different forms of reeds. And the idea was to see which one grew best.'

He got a strong clue from the name of one of the farms in the estate holding: Little Carr Farm. 'Carr', spelled with a double 'R', is Old English for wet woodland, suggesting strongly that is what flourished here before drainage and farming came along. James checked with scientists that trees that thrive with wet feet would still protect the peat, and they agreed. But what about the food, I ask? He replies: 'I know I can turn wood into energy because it's been done for thousands of years, and I know I can turn energy into food with indoor farming.'

James is a big fan of farming inside – be it under glass or lit by LEDs in a warehouse – not least as it helps control two things that he finds most unpredictable: weather and labour. 'With controlled environment agriculture,' he says, 'I don't have the same climate problems that I do trying to grow a crop in the field. And if I've got year-round production, I don't have a seasonal labour problem.'

This system would largely arrest the carbon loss from the field while still producing some food, but James wants more. Could he rebuild the carbon store? The word from the scientists was that leaf litter and twigs might build up the soil by about 1 millimetre per year, he says: 'And I went, "Oh, so 4000 years from now we'll be back to where we started from before we drained the peat." That's not really good enough. Hence why we are about to build a pyrolysis plant.'

He won't be just burning wood in a furnace to make heat and power, he'll use a pyrolizer. Pyrolysis is essentially heating something without oxygen, like the process of making charcoal. The plan is to coppice the woody plants – willow, alder or something more reedy – and feed that into the oven.

'At 850 Celsius we will turn the wood into solid carbon, biochar or engineered coal if you like. We then bury that underground, hence the name of "Reverse Coal" for our project. Making biochar gives me renewable electricity and heat, but some carbon dioxide is given off, which I will then pump into the buildings for indoor growth. Plants grow faster in higher concentrations of CO_2.'

The biochar will break down eventually but it takes many centuries depending on conditions (a longer duration than simply tree-planting). In the meantime it is holding the CO_2 and helping the plants to grow. Biochar is not a fertiliser itself but it's an unusually welcome home for microbes. All the tiny pore spaces that once made up the cells and channels of the living plant are great shelter for soil bacteria. It has been calculated that a fistful of biochar has the internal area of a basketball court.

The same patch of land will produce food, energy, carbon storage and habitat. That is a towering stack, but James isn't done yet.

'We've got two rivers that flow through the farm: the River Idle and the River Torne. Both those rivers have flooded and further upstream are Worksop, Retford, Doncaster, which have all flooded. So, if I now grow a crop that can sit under water, I can make my land available as a floodplain. I can do that. Secondly, both those rivers are failing the water framework directive [a pollution regulation]. Well, I can grow things that can improve water quality, like reed and willow which can strip nitrate and phosphate out of water. So I can improve the water quality.'

If James delivers, this might well be the most multi-functional land I encounter. The pyrolysis plant is due to be completed in the summer of 2024, around the same time that he will have a couple of years of data from his trial plots. Armed with those results, he then plans to plant up 250 hectares of his farm with a mosaic of the best performers. But he also has plans beyond the boundary, having identified 35,000 hectares of the Humber-Head levels that would be suitable for similar transformation. With a pyrolysis plant of a size to match, he reckons the combination of avoided emissions from degrading peat and stored carbon in biochar could save 1 million tonnes of CO_2 while still stocking the vegetable aisle. But why stop there?

'In total,' he says, 'there is a possibility of ten more in the UK of a similar size. Lapwing Estate has also identified similar opportunities in Belgium, the Netherlands, California and Malaysia.'

I'm reeling and have the little 'too good to be true' warning light flashing. Does the science back the vision or is this just agricultural 'cakeism' – eating it, yet still having it? The Centre for Ecology and Hydrology have their monitors bristling across many parts of the farm, measuring the exchange of gases above the different land uses beneath, and Chris Evans is cautiously optimistic.

'I really like James' vision,' he tells me, 'and there are no logical or scientific flaws, that I can see, but quite a few practical ones. At the moment we struggle to keep fields wet as, despite the fact that they are below the river, they are surrounded by drained fields [on neighbouring conventional farms] and the water runs to the lowest place. Bigger scale will help with this. The economics of energy for indoor farming are quite tricky and investors are wary of something so radical. So it will be tough to deliver, but James

Brown is a visionary and we [at the Centre for Ecology and Hydrology] want to make it happen in real life.'

That's a pretty positive endorsement, but Chris did have one reservation, which I'd raised myself on the tour: given we know that solar PV makes around fifty times as much energy from a given area than a crop, why not plant panels? James is planning to try out some solar on rewetted peat but he doesn't think it's the whole answer. It doesn't deliver material to make biochar to boost soil carbon, but also he needs reliable, year-round power for his indoor farming – something to keep the plants warm when the sun doesn't shine: something like a fuel.

James thinks land can and must do many things at the same time. But, for someone with such deep-held environmental beliefs, he is unusually passionate about food production. He condemns off-shoring our national ingredients, and says he sees it as not just a carbon dodge but a security risk too.

'My title of my dissertation for the International Relations course was "Is there a moral case for subsidising food production?" We obviously spend money on defence, we spend money on health. If you don't have food, none of those things are actually worth anything. What is the point of defending us if the country can't feed itself? And actually, it is more fundamental than education, it is more fundamental than health, it's more fundamental than defence, and yet we leave it to the free market to provide entirely. It is not defined as a public good but it's a fundamental piece and it should be classified as a public good.'

As I'm about to leave Lapwing Estate, James beckons me up a modest slope to present the 'after-dinner mint' following his banquet of land uses.

'This is a 30-million-gallon reservoir. We can run fifteen irrigators from the system. It's about four hectares of water. And floating in the middle of it are solar panels, 400 kilowatts. We've taken land out of production for water security but then what we've done on top is layered solar panels which float up and down with the water. They power the irrigation system, packhouse, cold stores and the office, which is behind us.'

As it becomes increasingly accepted that 'business as usual' in Fen farming is losing its grip, the wildlife groups spot a huge opportunity to move in. Inland wet and wild habitats have been reduced to small pockets, and wetlands still lack the popularity of woodlands. Marshes have long had an image problem: not solid land or noble sea, they were seen as treacherous places where the unwary might meet a watery grave. In hot climates they could harbour clouds of mosquitos and spread disease from the malarial swamp. Personally, I find these liminal places invigorating – they offer an environment beyond certainty, where you can't quite trust the ground beneath your feet, and for me that's fun. One such spot that makes me happy is Woodwalton Fen in north Cambridgeshire. It is one of the last remaining scraps of original Fenland, saved as a wildlife reserve in 1910 by the banker and conservation pioneer Charles Rothschild. It is a maze of reedbeds and bullrushes, dykes and ditches, scrub thorn bushes and toppling willows. Nine hundred species of moths have been recorded there, while kingfishers streak the waterways and marsh harriers hover above. But what's really impressive about this place is what it's becoming: part of the 'Great Fen', a project to recreate the areas original wetlands across 3700 hectares, thanks to a partnership between the Wildlife Trust, Natural England, the Environment Agency and the local council. Funded by a mix of lottery money, government funding and donations, they have been steadily buying up neighbouring farms. Many of those holdings boasted a good harvest and most of the land will be managed for nature recovery, so the 'farming vs wildlife and carbon' debate is unavoidable. Is there a solution? Step forward paludiculture – the growing of crops on rewetted peat, aka 'bog farming'.

Paludiculture is being trialled in Germany and rewetted parts of Indonesia, while the Wildlife Trust are leading the UK's first field-scale experiments. Crops they are working with include sphagnum moss (naturally antiseptic sponge for use as wound dressings or sanitary products), bullrush (the fluffy seeds make a down substitute for clothing and bedding, while the stalks can be chopped and pressed into insulation boards for the home), reed (for thatching and basketry), sweet manna grass (a wetland cereal crop), and water mint, wild celery, meadowsweet, cuckoo flower, hemp agrimony and watercress. The UK central government has stumped up £5 million to help with commercial development as, in their own words: 'with increasing concern about national food security, the need to develop paludicultural food crops is rising up the agenda'.

However, we are in the very early stages of experimenting with plants that can grow with water around their ankles. If

humans find marshes difficult to navigate, think how tricky
it will be to design farm machinery that doesn't get bogged
down. Most of the proposed plants are either exotic food
crops or novel commercial materials, which won't help to
fill the gap left in the nation's breadbasket from discon-
tinuing traditional farming on the land. The government's
own document goes on to dampen but not crush expectation: 'The opportunities are exciting and the potential
environmental gains to be realised are large. However, the
commercial growing of these crops in the UK is at a very
early stage. There are many barriers to be overcome on how
to sow, grow and harvest these crops as well as a need to
develop the market for these products.'

Yet it would be wrong to dismiss something so new just
because it cannot immediately compete with farming systems
that have grown up over centuries. Given the importance of
restoring peatlands to slow climate change, the fact that this
new way of farming offers some commercial value to these
lands makes it a necessity-driven innovation. Paludiculture
is multifunctional: it enables farming, carbon storage and
enhanced wildlife on the same plot while – for now at least
– shifting growing food from being 'king of the heap' to a
bit-part player.

The conflict between peat and farming is not limited
to temperate zones. Some of the deepest peat seams,
stretching down to 18 metres, are found in the tropics
and are composed mainly of decaying leaves, roots and
branches of swampy forests. Sadly, many of these have
been drained for farming, particularly oil palm plantations,
and the global-warming effect can be even swifter as hot

temperatures encourage faster oxidation. Wildfire is a risk too, as we know that peat can burn readily – and once did widely in the grates of Ireland and the Scottish Highlands. In 2015, a huge wildfire in Indonesia burned 2.6 million hectares across the archipelago, an area a little bigger than Wales. It released 2 million tonnes of CO_2, months of toxic air pollution, cost the country billions of dollars and was largely blamed on degraded and dried peat. However, this acute disaster had an impact on the public conscious-ness that the decades of emissions from steadily drying peat had not. You might even say it had a silver lining: the Indonesian government got serious about rewetting peat and, with the help of international agencies, claims to have rewetted 3.7 million hectares. NGOs and investiga-tors put this closer to 2.7 million hectares. Now, you can decide whether that is a glass nearly half full or more than half empty, but it is an area more than twice the size of the entire UK Fenland and this probably would not have happened without the catastrophic fire. Interestingly, the areas of least success with rewetting are those granted as concessions for logging or oil palm plantations – a striking example of commodity production triumphing over carbon conservation in land use.

There is no doubt that what to do about farming on peatland is a real puzzler, but I think we have to search for ways of weaning ourselves off it – not least as it is vanishing anyway. I wonder if a carbon strategy for all of agriculture could help. After all, if we look beyond peat, typical arable farms do have the potential, when managed smartly, to sequester rather than emit CO_2 and we've already seen land managers aiming for that. Indeed, there is a big buzz in farming over being paid via carbon credits for managing your land as a carbon sponge. The source of that money is the European Union Emissions Trading Scheme (ETS), which obliges some sections of the economy – power generation, heavy industry, manufacturing and transport – to pay for their carbon emissions. Farming is exempt. If it weren't exempt, peatland farming would be much less attractive. The cost of emitting a tonne of CO_2 on the ETS varies as it is a trading scheme but, even at the lowest reserve price of £22, this would add up to a surcharge of around £500 for farming a hectare of peatland. The phased introduction of such a scheme would drive climate-smart solutions on peatland or drive farmers off it, but it would only be effective if we could guarantee that any consequent drop in food production wasn't replaced by high-carbon imports.

7

WOODLAND

I'm bouncing along a forest track in what the driver calls a mule, which is like a rapid off-road golf buggy. It's a bright autumn day and the air is speckled with insects, many of which I am eating.

'I don't really like things I don't need and that includes a windscreen,' says Ellinor Dobie, woodland owner, forester and, at this moment, my driver. 'In the winter I wear ski goggles. Keeping the vehicle light is critical to getting round the woods.'

Ellinor and her father, William, have 278 hectares of forest around the Scottish village of Abbey St Bathans, about an hour south-east of Edinburgh. It's mostly commercial conifer stands but includes nearly 40 hectares of broadleaf trees – some ancient, some modern. It has two essential functions: making money and making habitat.

About 31 per cent of global land area is tree-covered and that figure has dropped a little over 1 per cent in the last thirty years. Forest cover is declining faster in tropical regions but is expanding in much of the Northern Hemisphere (though there is some concern this trend for more temperate forest might be undercut by the increasing numbers of wildfires driven by climate change). Russia, Brazil and Canada have the most trees, in that order. Australia makes a surprising entry (to me, at least) at number six, well above Indonesia.

I love trees and timber in every form. As a child I loved climbing them, especially a Bramley apple tree in the garden that had the plumpest fruits on the highest branches. My favourite room is pine-panelled. I'm a happy but bodgy carpenter and I love the work that goes into largely heating my home with logs. A few years back, even my TV publicity pictures had a lumberjack vibe: all chequered shirt and woodpiles. This love of trees is quite common, at least in a milder form. It's acknowledged in most hearts and heads that trees are 'a good thing'. They produce oxygen, store carbon both above and below ground, regulate the weather, clean water, house so much wildlife, provide solace for the mind and, in many parts of the world, are home to indigenous communities. How's that for multifunctional?

'Save the Rainforest' was one of the founding environmental campaigns and you'd be hard-pressed to find an individual or politician who is proudly anti-tree. But all those good things listed above don't generate wealth in a typical economic model. They are ecosystem services – something we increasingly recognise but still fail to value – whereas removing the tree, selling its timber and converting the land where it stood to grow food, makes money and creates growth. It's a well-worn but instructive truism that the GDP of an economy based on timber will carry on growing until the last tree is felled. And then what? This is a pretty good allegory for our exploitation of natural resources.

The focus of this chapter will be on what we can do at

home rather than in tropical jungles. This is partly because there isn't a huge dispute over how we should use rainforests: just leave them alone for the sake of the people who live there and the life-giving natural services they deliver; we should try to create political, economic and legal frameworks that enable this to happen. It is widely accepted that, while they sometimes work hand in hand, farming is a far bigger driver of deforestation than timber extraction. Some conservationists – notably the American organisation Nature Conservancy – argue that a form of benign timber exploitation called reduced-impact logging should be encouraged. The idea is that you select and remove individual high-value trees without clear-felling whole areas. This gives the forest an economic value, which is important for preservation (whether you like it or not) as it makes it less likely to be destroyed for farming. (For further detail, see my previous book, *39 Ways to Save the Planet*, in which I dedicated a chapter to this topic.) In looking at woodlands closer to home, I am going to raise similar questions: when can the chainsaw be the forest's friend? Is total protection the key to long-term survival? How do we achieve the best balance between exploitation and conservation?

Since 1990 the forested area of the twenty-seven countries in the European Union increased by 14 million hectares – 10 per cent more – almost equivalent to the land area of Austria, Belgium and the Netherlands combined. Yet in the same period, the EU increased its commercial timber production by 29 per cent. This is achieved partly by regulation (for instance, in Sweden you must plant two trees for every one you fell) and innovation that enables more of the tree to be used. In some places, especially parts of southern and eastern Europe, trees have naturally spread into abandoned agricultural lands. This productivity puts the EU in the remarkable position of providing over 40 per cent of global wood products by export value while having only 4 per cent of the world's forests.

The picture for life within Europe's forests hasn't been so rosy, as the planted areas have often been conifer monocultures, but in recent years this too has improved as governments insist on greater species diversity in order to qualify for grants. For instance, the UK Forestry's latest standard insists that there must be a minimum of two species in a 'compartment' (a stand of trees on similar soil bounded by tracks or firebreaks) and no one species can cover more than 65 per cent. Furthermore, native trees and shrubs should make up from 5 to 15 per cent. Both these requirements aim to help biodiversity, reduce disease risk and vulnerability to extreme weather.

Back to village of Abbey St Bathans. Tucked in a tight winding valley with a single road and the river Whiteadder Water running through it, it has a *Lost World* quality (or *Brigadoon*, for those with a knowledge of 1950s romantic musicals) and feels a little separated in time and place from the UK of today, but the forestry practice here is cutting-edge. In partnership with her father, Ellinor Dobie is transforming the estate management towards Continuous Cover Forestry (CCF), which she tells me 'is where you're trying to keep the forest environment intact in perpetuity whilst extracting timber. It isn't clear-fell rotation, which is the dominant practice in the UK.'

Clear-felling has happened when you see a swathe of land that looks like the aftermath of the battle of the Somme, with stumps, twigs and a few spindly stragglers remaining. The trees have been harvested like a crop, leaving woody stubble behind. The machines that do the work are called harvesters. They roll through the forest, grabbing trees in a giant metal fist before slicing through the base like it was a dandelion stalk and rolling the trunk up and down through its 'fingers', stripping all the side branches. The trunk is then cut into equal lengths. All done in less than two minutes. It's brutal but quick and cost-effective if you have large fields of identical trees. However, for the other life forms that lived there, it's a disastrous moment as the environment changes totally and suddenly. The choice is move or die. Some regeneration and pioneer species can enter the void quite swiftly, but it is widely recognised that clear-cutting is not conducive to a vibrant forest ecosystem.

Ellinor has driven me up a forest track to a point where she can show me how Continuous Cover Forestry would work. She stops beside a steep slope and we peer down through the rough trunks.

'Here, I'd want to enhance that promising-looking Douglas fir. So we'd clear some of the space around it and take out some competitor trees. These aren't wasted but of a lower value. This "halo thinning" allows the good trees to gain much greater value as strong construction timber. Some natural regeneration will come into that space but when we cut a big tree down we'll probably do some deliberate planting.'

More fundamentalist CCF practitioners think you shouldn't plant but just let nature decide what grows. Ellinor's not buying it, though.

'Some people think regeneration is almost sacred,' she says, 'like it wants to be here. But none of these trees here are native, not even the Scots pine, particularly for this region. I don't know where the seeds that take root here were blown in from. Anything that blows in and regenerates isn't necessarily well suited. It's just what's got here first. I would want to plant a great range of species, and even some of the same species but from different provinces. I think that results in more diversity and more resilience in the system.'

The end goal is a patch of land that is delivering both timber and natural habitat for ages, she says.

'What you're trying to get to is a forest environment with mixed age, mixed species. And so that will mean that it can continually regenerate. You're continually getting young trees; you're continually harvesting mature trees, or some you just let go into senescence and die. You've got a mix of different species, so they can all find different niches to regenerate and to grow. And then that's giving you resilience as well. You're more likely to have a forest environment into the future because if anything gets hit by disease there's plenty of other species to take over.'

'Resilience' here refers to the greater strength of a varied population to resist adversity – be that disease, storm, drought or climate change more widely – and, as our concern about those four horsemen rises, so does the clamour for resilience. Yet Ellinor loves the abundant nature for its own sake, she tells me: 'I live here in amongst everything else. And as well as wanting to stay living here, I want everything else that lives here to stay living here.' She believes that variety is the key.

'With a lot of different species, you're getting a much greater variety of niches. Different barks will be good for different lichens, different seeds will be good for different birds, and different mycorrhizal connections in the soil will be good for different fungi and different bacteria. That sheer diversity is good. Then also you're creating structural diversity as well because of these gaps that you're creating for your regeneration. That's also letting in light, and then you're letting in a whole load of different vegetation. You've also got age diversity as you've got "frame" trees that you'll need for a very long time and they'll become really mature and eventually get cavities and things that like living in them. And right beside it you'll have the very young trees and the mature ones and everything in between.'

We scrabble up the other side of the slope in among some trees that are just a few years old, their soft pine needles forming forgiving shrubby curtains up to head height. We push through to a clearing and Ellinor admires her brood: 'I think they're sweet. Cute. Adorable. I feel quite maternal towards them,' though not so maternal as to prevent choosing which of her 'offspring' is for the chop: 'We'll be selecting the best trees on form as well as on species and felling out their main competitors and then we'll do that repeatedly for the next five years.'

I can see the advantage for nature but, when Ellinor describes the process of CCF, especially on a steep slope like this, it sounds expensive and perilous. Corridors are cut so that trees fall into a gap and allow the extraction of timber (usually by winch due to vehicle inaccessibility and to avoid greater soil damage) and most of the harvesting is done by a lumberjack or lumberjill with a chain saw.

'On the face of it,' Ellinor says, 'clear-cutting seems a lot cheaper and a lot more efficient because you can go in with massive machines and get it done extremely quickly and get it away, and you get your income and it looks very impressive. But a huge chunk of your income is going to go on fencing [to allow regrowth after harvest] and restocking, so then it starts to look a lot less impressive, but a lot of people maybe don't quite put those two things together and view them separately. And you're also getting risk of disease and risk of drought.'

The real asset for CCF is a small local sawmill, especially if you own it yourself. The estate that Ellinor manages has had a sawmill since the mid-1800s and the old water-powered table saw is still there in a weed-besieged shed beside the river. Ellinor grew up around the modern sawmill and can operate most of its machines. It's spread through a handful of open-sided buildings, each one surrounded by timber in various states of preparation: tree trunks, planks, beams, stakes, panels, decking and woodchip. There's a truck with a trailer that looks like a giant upturned metal rib-cage on wheels and a tractor that's been buckled by a falling tree. Most of the timber left over after milling construction materials goes into a boiler to heat all the homes in the village, including Ellinor's. It's a mini district heating system with the hot water flowing to the various houses through heavily insulated pipes.

The sawmill itself can process about 1000 cubic metres a year, made up of 80 per cent pine and 20 per cent hardwood. The price of timber varies hugely year on year, making economic survival precarious, but it's not just about the money, it's key to Ellinor's forestry philosophy: 'Continuous Cover Forestry fits the needs of small and medium sawmills because you're getting a slow and steady stream of timber from a site rather than one massive influx. You can also welcome smaller batches of much higher-quality timber for use in construction.' She says the sawmill mirrors natural cycles: just as when big old trees eventually tumble and allow space for young upstarts, they cut down the big trees, which hold the most value for the least effort.

'I think to be ecologically, environmentally sustainable and biodiverse, you have to mimic nature,' she says. 'If you have a forest environment that has a mix of age and has a mix of species, as it would naturally, you also need a scale of disturbance that would happen in a natural system. You want to keep the scale of disturbance relatively small – smaller than is currently done in the UK [with clear-felling]. To do that in the current way the industry is set up can be very challenging. There are lots of foresters I know who would love to plant a wide variety of species, who would love to do something like this, but it's just not economical when it's going into large sawmills that just want a huge quantity of a certain species of a certain size. So it can be very difficult to do what a lot of us would perceive to be ecologically sound forestry.'

Before we return to the woods Ellinor responds to my prompts about being a rare thing herself, a woman in forestry.

'Sometimes that's a good thing as, starting out, people remember you. People invite you to things. You often get more opportunities because you stand out. But in other ways, there are maybe microaggressions or people put you down or you wouldn't be chosen to do things. Or people will assume that the man has more knowledge and so choose them over you, or you need a lot more knowledge than them to be seen as being professional. Or whenever you do something or say something, you feel that you have to be really good at it because you're representing all women. So that's a huge amount of pressure and often I'll back out of that. I can easily load up anyone's trailer with a forklift. But if there's a guy there, I will definitely be like, "Oh well, he can do it," because if I do it wrong, I might feel I've let down all women.'

For the vast majority of the time, though, whether among the blades in the mill or the trunks in the woods, Ellinor says she just gets on with the job, regardless of her gender.

Much of the Dobie family woodland cloaks the steep sides of the valley so, as you meander beside the river Whiteadder Water, the forest is on display. However, that cloak has recently been shredded. On the night of 26 November 2021, Storm Arwen, armed with winds around 100 mph, flattened one quarter of their forest – 70 hectares of woodland, or 25,000 cubic metres of timber – enough to keep their small sawmill fed for twenty-five years. Across Scotland the figure was 4000 hectares, with 1 million cubic metres of timber – enough to fill 400 Olympic swimming pools. Three people lost their lives. The destruction extended down into north-east England. Ellinor slept through the storm but, days later, what happened that night brought her to tears: 'When I went to the yard in the morning, we knew something extraordinary had happened, but it was weird. Maybe it was nerves, but we were laughing too. It took a few days to appreciate the scale of destruction.'

This is the second time I've see Arwen's handiwork, as a couple of years ago I made a radio documentary about its wider impact and recovery plans. Both times I have been struck by two things. The first is the patchiness of the destruction: perfectly intact forest next to a massive hole. It's like a giant boot hundreds of metres across had stamped here and there. The second is the apparent ferocity of assault in those patches: thick trunks splintered like matchwood, and massive root balls flung around like toys abused by a toddler in a strop. As Ellinor points out, this is clear-felling by mother nature and it tore up her gentle, caring forestry plans. They will still extract the fallen high-value timber like larch and Norway spruce for their own sawmill, but the Sitka spruce and Scots pine that were toppled on the flatter ground is being taken by a contractor and sent off to a large sawmill (although even they are only removing a small amount as they're swamped and so the price has dropped too). Some of Storm Arwen's holes in the fabric of the forest have already been cleared, and Ellinor sees this as an opportunity to start CCF from scratch, a blank slate.

The first thing I notice in one of these wind-scoured clearings is that it is now very well protected with 2-metre deer fencing double-wired over the bottom half with mesh to keep out hares. Ellinor makes sure I close the gate in case a pesky chancer sneaks in behind us; both species like nothing more than nibbling a tender sapling: 'The trees outside the fence have been dipped in Trico, which is an emulsion made of sheep fat and a natural deer deterrent – so far it is working well.' Most of the established forest is not fenced so deer are controlled by shooting, which has the collateral benefit of supplying one of the healthiest and morally defensible meats (declaration of interest: Ellinor gave me venison jerky for lunch).

This newly planted 3-hectare block is at the base of the slope we were on earlier, looking at how to practice CCF, where it flattens out towards the river. When the storm struck, it was mainly Sitka spruce but Ellinor has planted it up with a range of species to deliver wind resistance, long-term accessibility and good biodiversity. Around the outside is oak, birch, hornbeam and sycamore to provide a permanent windbreak on the edge. It is divided periodically with rows of Norway spruce to allow extraction tracks in the future. The sunniest and most sheltered spot has Douglas fir and wild cherry. We are walking through crunchy grey twigs, which is what remains of the brash from the fallen trees, regularly punctuated by various saplings ranging from knee high to waist high. They only went in six months ago, mainly planted by Ellinor.

'These mixes have been designed so that the different species nurse and help each other,' she says, 'and also reach maturity at different ages so that the faster-maturing species will be reduced in number sooner, leaving space for the longer-lived and more valuable species to put on girth; both will also then have the space to regenerate.'

Her enthusiasm is infectious and she seems totally unfazed by forestry being the very definition of delayed gratification. The fruits (or logs) of your labour won't be seen for decades, in some cases not even in your lifetime. But, for Ellinor, as it's not all about the timber yield, she can also enjoy the habitat she is creating, and that brings rewards every year, if not every day: 'We are already thinning trees that were planted the year I was born, twenty-eight years ago, and they are already of some use. My dad, who's sixty-eight, is milling trees into good timber that were saplings he weeded round as a boy. It's true that I may not see the fully regenerated mature forest environment. I'll be dead by the time there are owls nesting in the old trunks.'

There are plenty of twisted gnarly trunks in the 22 hectares of ancient woodland we visit next via a footbridge across the tea-coloured waters of the river Whiteadder Water. Here again the trees clad the slopes. These are nearly all oaks of around 200 or 300 years old but, despite their age, they are not tall. They are stooped and twisted, with crooked limbs at inexplicable right angles, and are all the more interesting for it. Thin soil and a windy location doesn't favour tall trees. Two thirds of the woodland is designated as a Site of Special Scientific Interest and the presumption is 'do not touch', but Ellinor's expert eye spots problems that could be lessened with a bit of human intervention: 'There are no young ones. There are no teenage ones. They're all mature and going into decline or senescence, all of them.' Without younger stock to replenish the forest, it's not truly sustainable. Shading and browsing is stunting the youth and has done so for decades. I point to a baby oak reaching up to my calf and wonder if that is a sign of hope.

'We're getting small regeneration,' Ellinor replies, 'but it can't get any higher because it's just not got enough light. And while it sits there waiting for light, it gets browsed by the deer and the hares and the rabbits.'

Ellinor wants to lop off a few old limbs or even cut down a handful of trees, put some deer netting around the gap and see what grows. They might even find a very satisfactory use for the timber: 'We have had a couple of boat builders get very excited about crooks, where the grain of the tree is following the curves they need on a boat. It's a heck of a lot stronger if the grain follows the curve rather than joining two pieces. But it's a very niche market.'

The level of protection here means that any management needs the say-so of officials, in this case NatureScot, Scotland's nature agency who pledge to 'maintain, enhance and share the benefits' of the natural environment. Ellinor understands their caution. Neither I nor Ellinor are saying that CCF is always the best approach to commercial wood production, and she believes there are some geographies and market pressures that might necessitate clear-felling. Clear-cut plots always look terrible from a distance as all you can see are the fallen and the dead, but often – once you walk within them, even quite soon after harvest – you can see new life emerging. What draws me to Ellinor's method is the sheer intensity of thought and management: the granularity of looking at virtually every square metre in terms of fertility, habitat, light, windage, neighbouring species, future access and much more besides before deciding what to do with it. The result is truly multifunctional, delivering a crop, a home for nature, flood protection, beauty, recreation (there are footpaths and mountain bike tracks) and carbon storage.

For many people spurred on by government targets or a belief that tree-planting can offset their emissions elsewhere, carbon storage is a tree's principal function, but Ellinor believes this singular obsession can be harmful.

'Trees are not just carbon sticks,' she says. 'Sometimes they [carbon counters] oversimplify or even sort of compromise nature. Nature doesn't go with what we think is good or bad. The forest doesn't care about carbon. So trying to always get the carbon argument into the forest is quite difficult because it's not its end goal to store carbon. If you want to just absorb carbon quickly, plant Sitka spruce close together, grow it fast and harvest it young. That's not the best for biodiversity or ecology, but that would get the most timber, most carbon into timber, into buildings and sequestered until the building decays. But I think it's not that much carbon in the grand scheme of things. So you're doing a lot of damage for an unnoticeable difference.'

She also points to conflicting priorities: 'In the long term, mixed resilient forest is best all round, but for the next few years with the climate emergency, you could make a case for it [intensive forestry].' Nevertheless, with a mixture of management approaches, Ellinor thinks there is more room for much more tree cover in the uplands of Britain. When I ask her how she feels about some of the broad barren hillsides in Scotland, I'm impressed by her powerful hair-care metaphor: 'It's like they've been waxed and it reminds me of that pain.'

Returning to the carbon story of trees, there are two parts. There is carbon *capture* as the tree grows and carbon *storage* if you use the timber for something long-lasting. The desk I am writing on, the floor it's resting on, the elm I salvaged from a nearby copse to make a support pillar – all are carbon capture and storage. Go further back and our churches, castles and country houses store carbon in their timbers from the date they were built. New innovations – especially cross-laminated timber, which is like slabs of chunky plywood, enabling the construction of walls, floors and lift shafts – mean that wood is becoming a fashionable building material once again. Its other asset is what it is often used instead of: cement and steel, which both emit huge amounts of CO_2 in their production. It's estimated that close to 14 per cent of all the world's human-made CO_2 emissions comes from the manufacture of those two materials.

I was so convinced by the virtuous climate credentials of wood that it made a chapter of *39 Ways to Save the Planet* and included one of my favourite statistics: every seven seconds the sustainable forests of Europe yield enough wood to build a four-person family home. I give talks where, with the help of a hydraulic press, a steel bar and a wooden length of 'two by four', I demonstrate that wood is stronger than steel, weight for weight. So I was surprised and a little shaken to read the following headline in July 2023: 'Wood is not the climate friendly building material some claim it to be'. And this wasn't from some industry-funded think tank but from the World Resources Institute (WRI), who I have trusted and quoted liberally elsewhere in this book. Given the statistics at the beginning of this chapter, the experience of foresters like Ellinor Dobie, and the comparative 'evil' of other building blocks, what is the WRI's beef with wood?

They have two key arguments. Firstly, they point out that far from all of a tree is used in the final timber. The bark, branches and root mass are discarded to rot or burn – both sources of CO_2 often not included in the carbon life-cycle analysis of construction timber – although they concede that some of these by-products are used to make paper and chipboard. Secondly, they argue that most calculations ignore the carbon that would be sucked up if you didn't harvest and allowed the forest to grow. They include both tropical and temperate forests in their calculations, and they also cover both felling for firewood and building.

This is really interesting so let's unpack it a bit. Normally, 'sustainable harvesting' (as frequently trumpeted in Europe and North America) means taking no more from the forest than it regrows in the same time period. If that regrowth means as much or more carbon is held by the same area of forest, then the impact is said to be carbon-neutral or even carbon-positive. But the WRI argues that we should be comparing timber extraction to a 'no-harvest scenario', where the managed forests were left alone to grow, not just tick over, as growing forests are carbon sponges. They use the analogy of a bank account gaining interest. The harvesters are like someone taking out money equal to that interest all the time and saying, 'Look, I'm having no impact on money in the bank.' By contrast, 'no harvesting' leaves the interest in place and the stored money swells. More money is held when you don't cream off the interest; more carbon is held when you don't fell the trees.

The WRI's final projection is that forest harvesting has a carbon *cost* of 3–4 gigatonnes per year, even considering the benefit of substituting steel and cement. That's approaching 10 per cent of all anthropogenic emissions – slightly more than all transport. They argue that this figure will get worse if, as expected and desired by some, timber harvests increase. But they also present this as an opportunity.

'These findings are, in a sense, good news because they imply that if people could reduce forest harvests, forest growth could do more to reduce atmospheric carbon, a potential mitigation "wedge" that is rarely identified in climate strategies. As with other mitigation efforts, reductions have value only to the extent they do not shift emissions to another source. Over time, if more forests were able to mature this net sink would decline but these efforts would help "buy time" for more climate mitigation activities to become viable.'

In other words: using *less* wood for making stuff for the next fifty years or so would really help our immediate climate crisis. Time to try that out on some of my timber cheerleaders.

Michael Ramage is a professor of Architecture and Engineering at the University of Cambridge who is passionate about the climate and lifestyle benefits of building with timber. He believes people work and learn better in wooden rather than concrete rooms. He is '100 per cent behind' the idea that we shouldn't harvest timber to burn it. But that's where his agreement with the WRI's timber analysis stops. He doesn't trust the computer model they used to calculate future carbon emissions, and he doesn't think they should conflate tropical with temperate forestry regimes or firewood with construction timber. However, his main criticism is over what would happen to non-harvested trees: 'Most harvesters fell trees just past the peak of growth as older trees are better for construction and therefore more valuable. And trees slow down their growth rate radically after maturity, just like you or I. They don't store carbon nearly as rapidly as young trees planted in their place.'

Then he takes on the financial analogy: 'The bank account model is flawed. Forests are like a bank account that has interest but that interest is capped and slows towards that ceiling. Trees don't grow bigger forever. Let's say the account grows rapidly to £100 but has a limit at £110. Once you approach that limit and the growth rate begins to slow, you should take out some cash [timber] and store it in another account [a building] so your original bank account can return to rapid growth.'

He also points out that unmanaged temperate forests are becoming an increasing fire risk with climate change.

Mark Palahi is ex-CEO of the Circular Bioeconomy Alliance, which 'aims to accelerate the transition to a circular bioeconomy that is climate neutral, inclusive and prospers in harmony with nature'. He repeats the criticism above that most trees are harvested as their growth rate slows down whereas new trees are soon enjoying the spurt of youth. But the main flaw he cites is that the WRI doesn't consider economics, especially how demand for wood encourages more forests to be planted. He points out how most afforestation in both the tropics and temperate lands is done by those who want to harvest at some date and that these woodlands are much less likely to be destroyed for agriculture.

This analysis is shared by Mark Wishnie, chief sustainability officer of BTG Pactual, the largest investment bank in Latin America with interests in tree planting and harvesting. He is a forester and research scientist by training.

'The authors have made an assumption that economics does not affect land use,' he says. 'So [according to them] an increasing in consumption of timber equates to an increase in harvest. Full stop. But that is not how the world works. When you buy more of something it creates a price signal – you buy more wood, the lumber yard runs out of lumber, then prices start to go up. When prices for trees go up, people don't just harvest more trees, people plant more trees, people change forest management to produce different products, and mills and foresters waste less wood. Also, looking back, without demand those forests that have been planted would not have been planted. Demand for wood increases forest area.'

He reckons without demand, far from seeing increased growth, the forests could be destroyed, both in the tropics and in the north: 'In Brazil, there is about 10 million hectares of planted forest, most of which is planted on former cattle pastures. If there is no use for wood there is nothing stopping them going back to cattle. That's what we suspect would happen.' He continues:

'In the absence of management, forests don't grow, they disappear in forest fires. In North America and Canada, there is a desperate effort by NGOs and state governments to remove trees from over-dense forest where we have

suppressed fires and harvesting in the past. The holy grail
is a market that would pay for all that [excess] wood to
come out of the forest. There have been examples in the
last couple of years where fires have come through federally
protected land and had a devastating effect, but when they
come to privately managed land, the trees are more vigorous
and healthier, the stands are more open and that's where the
fire stops or can be stopped.'

He also thinks harvesting and replanting is a more effective carbon sponge than letting mature trees grow old:

'Without harvesting, forests get to a state of senescence. In
an actively managed forest you do harvest before that stasis
point. Forests grow fast in early years. If you are keeping it
in that younger age class then you have fast-grown vigorous
trees. You take a proportion of that for long-lived wood
products, then you are getting that virtuous cycle of rapid
sequestration, storage of that carbon and then new rapid
sequestration when you replant. From a carbon dioxide-
absorbing machine perspective, it's very effective.'

Tim Searchinger remains unpersuaded, saying the timber
industry underestimates the amount more mature trees
would grow if they weren't harvested, before adding that the
best thing commercial forestry could do for climate change
in the next few decades would be to stop felling trees.

In October 2022, I found myself uttering meaningless
vowels as I headed towards a dry-stone wall at 105 mph.
Simultaneous terror and exhilaration had the adrenaline
taps gushing. I was trying to articulate something worth
broadcasting as this was part of an investigation into the

value of carbon offsetting. I was in a rally car being driven (thankfully) by the totally skilled John Marshall, owner of Beatson's Building company and sponsor of the Isle of Mull Rally. At 214 km (133 miles), it is the longest closed-road rally in the UK and has been running for more than fifty years. The competitors are a mixture of locals with souped-up classics – Ford Escorts, Minis, Subarus – and modern works cars from manufacturers like Hyundai and Peugeot. It brings thousands of visitors and millions of pounds to an island normally cherished for its tranquil, natural beauty. CO_2 is being pumped out for fun and in recent years they have decided to offset 70 tonnes of it, which they say will cover not only the race itself but the journeys of the visitors. For comparison, the typical UK resident is responsible for about seven tonnes of CO_2 per year, so ten times that.

The idea of carbon offsetting is that the emitter pays for something to happen that absorbs as much CO_2 as they have created. The arguments around it are, in essence, quite simple. The argument in favour runs that, as a key part of our economy or leisure, these emitting activities are going to happen anyway and so, while trying to reduce their pollution at source, it's a good idea to fund equivalent absorption elsewhere. The argument against says that these schemes excuse companies and governments from actually cutting their own CO_2 output and that frequently the projects are ineffective – that in climate terms, it is like buying permission to sin. I am not going to delve much deeper into this as I think elements of both sides are true; where you sit is almost a philosophical decision. What is beyond doubt is that the amount of money in carbon credits is huge and

many climate-friendly activities like tree-planting, wetland
creation, renewable energy production and enhanced
geological carbon storage have grown with that finance.

The Mull Rally partnered with Carbon Positive
Motorsport and they chose Highland Carbon to generate
the carbon credits (they don't like the term 'offsets'). The
company was founded by zoologist Richard Clarke and he
showed me where the trees paid for by the Mull Rally had
been planted. They didn't have a verified scheme on the
island itself, so it was near the Scottish borders as part of
a larger plantation rising up from a small loch: 'There is
broadleaf lower down with willow, alder, aspen and oak
then more conifer towards the top.' The saplings had just
gone in and some were so tiny that we had to watch our
step. 'Elsewhere, we plant rowan, hazel, birch and wild
cherry alongside the fast-growing timber species like Sitka
spruce and Douglas fir. As a zoologist,' Richard said, 'I
love some of those conifer species as they are great for red
squirrels and pine martens, but it is just not true that we are
just spreading conifer blankets.'

For any plantation to generate a carbon credit it must
follow the UK Woodland Carbon Code, which has rules
on tree variety, location of the new forest and its management. They have certified more than two thousand woods
in the UK. Aware of the need for trustworthiness, a whole
forest of verification schemes has taken root. Highland
Carbon's website alone has a banner of eight across
the top, including VCS, Gold Standard, the Peatland
Code, BioCarbon Registry and United Nations Clean
Development Mechanism. 'All our projects are third-party

inspected,' Richard told me. 'There are inspections after planting, after five years, and ten years after that. They are not just checking that they are there but that they're storing enough carbon.'

Richard believes that a forest financed by carbon credits won't look much different from others. In fact, it will look much better than the monoculture commercial plantations that sprouted in the Highlands in the late twentieth century. He also insists that both the buyers and sellers of carbon credits are not the money-grabbing or polluting stereotype.

'Most landowners I deal with are passionate about wildlife,' he said. 'I am selling them a whole narrative of how their landscape can be improved for climate, nature and flood reduction. If the narrative is strong, they will pay more. The other big misconceptions in the voluntary offsetting market is that the company seeking the offset hasn't cut their own emissions. I meet with them before they buy credits from us, I see their third-party carbon audits, which show me what they have done, and most of them are doing a lot. These credits aren't cheap so it is more cost-effective to reduce their own carbon first. We haven't dealt with a single company that isn't acting to cut their own carbon. The cynicism is misplaced.'

I am aware, of course, that putting a positive spin on offsetting is in Richard's commercial interest but it seems to me that – with robust regulation of how and where it is done – forests funded by carbon credits are a smart use of land. There are pitfalls, especially the danger of massive carbon leakage from the soil when woodlands are developed in the wrong places like peat bogs, but planting trees in areas of degraded land can be good for the climate, water storage, nature, recreation and, at a much later date, building materials.

Most of the tree-planting schemes delivered by philanthropists, fundraising or even private estates are not meant for chopping down. That isn't in the management plan; indeed, it might be seen as heresy. It is being done because trees are 'a good thing'. In Britain, we often self-flagellate over the fact that the UK (especially England) is one of the least-wooded countries in Europe, with only 13 per cent tree cover. We are thirty-eighth out of forty-four (above Ireland and Holland); Finland is top, with 74 per cent. But this UK figure was barely 5 per cent at the start of the twentieth century and has more than doubled in the last hundred years, mainly thanks to plantation forestry. Now the emphasis is on broadleaf trees, with habitat creation, recreation and carbon storage as the primary goals.

The National Forest – spanning 520 km^2 in the counties of Derbyshire, Leicestershire and Staffordshire – is an ambitious project to plant up about a quarter of it by encouraging and funding afforestation. So far, around 9 million trees have been planted. The purpose is as much about people as nature: these were lands physically scarred by mining and heavy industry, then socially scarred by their closure. The idea of the forest is to provide huge 'green lungs to breathe life into a landscape and transform communities'.

The Heart of England Forest is a quarter of the way to establishing 121 km^2 of broadleaf woodland in Warwickshire and Worcestershire. As part of their plan to tackle the climate emergency, Wales has a plan to plant 430 km^2 by

2030. Businesses, like the energy supply company OVO, have pledged to plant a million trees. The motorway and trunk-road operator, National Highways, has a commitment to plant an additional 3 million trees by 2030. However, I did a report for Sky News in early 2023, pointing out how hundreds of thousands of trees planted beside roads had died due to poor maintenance. If you want trees to make a difference to climate or nature, it's about *growing* not just planting. Even one of my own programmes – *Countryfile* on BBC One – launched a tree-planting campaign in 2020 that to date claims to have put 1,705,262 saplings in the soil. As a national licence fee-based broadcaster, the BBC can only campaign for things that gain almost universal approval: trees tick that box.

Not far from my home in the Midlands is a 70-hectare wood that has appeared in the last thirty years. Its bisected by a bridleway, which I bike through once every few years and on each visit have marvelled at its increasing beauty. But this clearly isn't the result of a simple rewilding project; the species variety, frequent ponds and broad rides mark it out as the work of a true landscape architect.

'If you look closely, on this aerial photo, you can make out the three letters of my wife's name in shape of the walk-ways between the trees. It's —.' I've left the name blank because the couple behind this wood asked for anonymity and I agreed because it fits with their motives for creating it in the first place: peace and seclusion. I only found them by cycling up the drive and asking. Let's call them Fred and Alice. They are not claiming it was driven by a great desire to address the climate or nature crises, but they have undoubtedly helped both. 'We simply wanted somewhere to live in a woodland,' they tell me. 'We couldn't find one we liked, so we created one ourselves. It's got a stillness that we like.'

They bought a fairly typical mixed farm: some cattle and 'acres of monocultures, with hedges removed in prior years'. They started with five blocks of rabbit-proof fence and planted trees direct into the cereal lines. The fertiliser gave the trees a good start and the growing crop kept weeds at bay. Their neighbours were puzzled, they say: 'Some of local farmers at the time said it was sacrilege to destroy food-growing land with trees. It wasn't antagonistic so much as perplexed. "Isn't the first important thing to feed people rather than for them to enjoy having trees?" I pointed out that the "Common Agricultural Policy" [the farming regime from the European Union] already demanded that some land should be "set aside" from food production, and trees were better than just weedkilling a patch.'

We hop in a buggy, similar to Ellinor's 'mule', and set off for a woodland safari. The first eye-catching vista as we drive away is back towards the house. It may be in the Midlands but it looks like the Canadian outback: they have totally succeeded in creating a cabin in the woods. Actually 'in the woods' makes it sound a little enclosed and creepy, whereas they have laid out meadow rides on 20 per cent of the land and dug twenty-six ponds. The area immediately around their home and outbuildings is very open. The motivation is a little like the great eighteenth- and nineteenth-century landscape architects who wanted to create appealing scenery.

They got grants to help finance some of the forestation, and manpower too, but Fred liked to get his hands dirty and deliberately subvert the regime. We jump off the buggy and head into a thick and rather damp clump, and Fred says: 'Professional planters would come through and then I'd plant extra ones to make it look more random. We also did it denser than recommended because then some die and you get further randomness. In my heyday, thirty years ago, I could plant a thousand trees a day. That was my record. Although when planting into ploughed tills, I used to break a lot of spades.' Decades on, you can see the evidence of his subversion: like a disorderly parade ground, the intended ranks are still visible but messy and, once you step away from the axis of the line, it looks pleasingly random. There is also such variety. Oak and ash are the staple ingredients of this forest but it's seasoned with so much else: poplar, Corsican pine, Christmas trees, silver birch, apple, pear, cherry, hornbeam, willow, field maple, walnut, lime, wellingtonia, alder and hazel.

Fred's cunning applied to shopping as well as planting, he tells me: 'I'd get trees cheap, or sometimes given them, at the end of the season because when the planting season is over nurseries want to switch off refrigeration. They used to burn them. So I planted them hurriedly in beds, tended them through the summer and dug them up and planted them in the wooded areas next season.'

Many of the trees are now 8–10 metres high, and the stands are big enough to lose sight of the edge and enjoy the frisson of losing your sense of direction. The sound is at once dampened but particular and the scents vary from the herby smell of the pines to earthy mould of the litter. There is a vogue for forest bathing now, where you rest in the woods and let the sensations wash over you. I can see the appeal, if not the slightly hippy-chic styling, and this place is certainly a plunge pool for the senses. But there are also innumerable views of the woods from the outside too. The broad meandering meadows provide framing for different glades. Just as garden designers talk about dividing your plot into 'rooms', where you discover different places as you move around, so it is here. Many of the more varied 'seasoning' species I referred to earlier are planted at the end of the rows to enhance the impression of diversity.

I ask Fred if he was tempted by rewilding to just leave a patch bare to see what appeared, and he replies: 'We didn't do it because we perceived the early stage as ugly. You get brambles and weeds, and very quickly it becomes inaccessible. We were also anxious to get it growing really quickly.' Thirty years on, the forest has matured, the owners have aged and attitudes have altered: 'Now our perception of ugliness has changed and we embrace it more than want to control it. We have let some areas take their own course – partly through choice but also not having time. I've gone into a "don't mind chaos" mode, whereas initially we were in parkland mode and I had wanted it to be like a big garden. Now I want it to be itself, to have its own character.'

That said, they do have one little accidental rewilding trial plot. It's an island about 12 metres across in the middle of one of the larger ponds, cut off from the attentions of large mammals – deer or humans – for decades. 'Nobody has set foot on there for twenty-five years,' Fred says. 'For years not much happened there apart from nesting swans. Then for many more years it was just a mass of brambles. Now it is completely wooded – abundant and overflowing.' It reminds me of the fertile outcrop in the Scottish loch I referred to in chapter 5. You can't see the actual ground of the island as the densely packed trees seek light and find it by dangling fronds over the water. It's like a giant floating and tumbling hanging basket. Alder and willow – both water-loving types – have triumphed here.

The patchwork of both open and tree-covered ground is revered by ecologists for offering greater diversity and abundance to plants and animals. It's known as the 'edge effect' or a 'mosaic habitat' when repeated over a wider area, and this is precisely what Fred and Alice have created here. Their initial desire might have been to create a perfect home for themselves, but the result has been a metropolis for nature. The ponds teem with waterfowl, grass snakes, newts, dragonflies, occasionally water voles and even ospreys stopping by on migration. There are goldfinches, king-fishers, wrens, robins, pheasants, doves, wagtails, kestrels, sparrowhawks and increasingly dominant buzzards, who love the rich diet of voles. There are also fallow, roe and muntjac deer – Fred reckons twenty or thirty resident in any one time. The woodland even helped 'revive' a locally extinct butterfly: 'We discovered we had purple emperor butterflies but the local museum said they became extinct in the 1940s and I said "they're not extinct 'cos we've got them" including one in my living room which had died.' He still has the persuasive evidence in a large matchbox, its colour mostly gone but with occasional glimmers in the right light: proof of a species recovery.

The emerging Eden I have outlined above is not free of anguish. In truth, as we move through his creation, Fred spends as much time seeing the faults as the virtues. And the most recent problem is the biggest: '60 per cent of what we planted were ash and I haven't found one that hasn't succumbed to ash dieback. The ash was my last bastion, it was a body blow.' It's plain to see as we drive around: standing tree shapes of skeletal grey or some infected but fighting. The fungus, Chalara fraxinea, blocks water transport through the branches, with the canopy dying first. As the tree struggles for photosynthesis, masses of leafy shoots sprout from lower down the trunk – so-called epiphytic growth – but this effort is usually in vain as the fungus spreads and the tree dies.

The dead wood is unwelcome but not useless: all the heat in their home now comes from log burners; their wood piles, drying under a redundant greenhouse-style swimming pool cover they got from a neighbour, are an enviable fuel source for a log-burner like me.

Fungus is attacking on another front too. They planted tens of thousands of poplars, especially on the southwestern side of the plot to provide a break from the prevailing wind. There are eight different varieties but within those groups many are clones as they are grown as cuttings from a single tree. They established well and grew fast, Fred tells me: 'But once they have grown to 6–8 inches diameter many succumb to rust.' This is a fungus that turns the leaves, yellow, brown and eventually black. 'It looks like autumn has come early,' he notes sadly. And that isn't the only evidence of harm: the ground is littered with broken trunks, and many poplars have lost at least their top third. 'The fungus damages the trees' capacity to deal with frost. It's like the natural antifreeze in the tree is gone and in the winter they split,' he says. He has tried planting new, supposedly rust-resistant, varieties but even these are showing ominous signs of infection. Unlike ash, poplar has a pretty low calorific value, so isn't favoured to heat the home. Thankfully, though, rotting wood is not wasted in a forest – indeed, it's critical to the circle of life.

There is yet another dislodged crown here in the form of the majestic oak beheaded by an invader: the grey squirrel. It's easy to become rather blasé about squirrel damage in our woods but I've spoken to many foresters who think our efforts to create more broadleaf woodland will be completely in vain if we don't drastically cull these parkland favourites. The Royal Forestry Society counts squirrels as a greater threat than disease or deer. Introduced into the UK in the late nineteenth century, there are plans to give them oral contraceptives and even ideas to introduce progressive infertility through what's called a 'gene-drive' to eradicate them entirely.

Fred notes that the squirrels didn't attack ash trees; oaks are their favourite: 'Before ash dieback arrived, the biggest single problem here has been squirrels that enjoy taking the tops out of oak trees. They nibble off the bark two thirds of the way up and the top dies and drops off. The tree is disfigured. The tree can't make its mind up if it's a bush or a tree 'cos the squirrels have taken the top out of it. We've got thousands like that.' This behaviour doesn't usually kill the oak, but bark is the tree's first line of defence and wounds caused by stripping can allow fatal pathogens easy access.

Given these assaults to three of his staple trees, you can understand why Fred's eye can frequently focus on the negative. He reckons his replanting efforts are even being thwarted by the deer, who seem to have got smart enough to push over tree guards to browse on the sweet sapling beneath. Nevertheless, overall the couple's bold ambition and energy have transformed a bit of basic farmland into a delightful woodland teeming with wildlife. Its precise make-up may alter in the face of disease threats, pest pressure or climate change but its evolving variety ensures resilience. This isn't an appropriate use for large proportions of farmland, but I welcome some large pockets of new woodland established just for the love of trees.

'Look, just over there,' says Alice in an urgent whisper. 'We should be able to get quite close in the buggy.' And, sure enough, we approach within 30 metres of a mixed group of fallow and roe deer about eight strong. It's a proper wildlife encounter, a lengthy moment of stillness as we all keep an eye on each other. Then, on an invisible signal, they slip away among the steadily growing trunks.

8

NATURE

How much land should we leave for other life on Earth and how can we share our own spaces better? This is the key question for this chapter and, some might say, for the plight of all life on Earth, including our own. At a fundamental level we need thriving plants and plankton to create oxygen, and much of our food production depends on other life forms – hence the phrase 'there's no life on a dead planet' and the warnings of a looming biodiversity crisis, which many feel deserves the same focus (and concerted international action) as the threat from climate change.

In December 2022 on the final day of the United Nations' Biodiversity Conference, a Global Biodiversity Framework was adopted that includes 'concrete measures to halt and reverse nature loss, including putting 30 per cent of the planet and 30 per cent of degraded ecosystems under protection by 2030'. An essential goal, it was agreed, because: 'The stakes could not be higher: the planet is experiencing a dangerous decline in nature as a result of human activity. It is experiencing its largest loss of life since the dinosaurs. One million plant and animal species are now threatened with extinction, many within decades.' The five direct drivers of this loss are changes in land and sea use, direct exploitation of other life forms (such as hunting and fishing), climate change, pollution and invasive species. The '30 by 30' commitment was adopted by 188 countries, including the UK, within their own borders.

So we've seen the problem, identified the causes and committed to solving it. Great! Of course, it's not that simple. These are promises not achievements, and the implementation of them is fraught with questions over who changes and who pays. But perhaps the biggest challenge is the apparent conflict between two apparent realities: one is the lived experience of billions of people on the planet; the other is the facts revealed by the scientists. As the human population has increased and those humans have consumed more fuel, calories and materials, our attack on other life and its consequent erosion has grown hugely.

Let's look at quantity first. Humans weigh ten times more than all the wild land mammals on the planet. Or, looked at the other way round, all the mammals – from lions to musk ox, elephants to rabbits – weigh just 10 per cent of all the men, women and children on this Earth. Add in our domesticated and farmed mammals, and the figures are even more alarming: according to 'The Global Mass of Wild Mammals' paper published in the US *Proceedings of the National Academy of Sciences*, pigs alone have twice the total biomass of wild mammals. And the wild mammals that do flourish are the ones that rub alongside us best like deer, boar or rats.

Now let's look at spatial dominance. Around 30 per cent of the world is still considered wilderness, and this figure is shrinking. The big chunks of wilderness are Antarctica, the Saharan and Australian deserts, and the forests of the Amazon, Russia and Canada, much of which are inhospitable anyway. As an interesting aside, many researchers think those northern 'boreal' forests are not strictly natural but the result of humans wiping out the mega herbivores, like mammoths, tens of thousands of years ago, which allowed many more trees to flourish.

Add in the pollution, to varying degrees, of our entire ocean with plastic, or the whole atmosphere with CO_2, and you get a pretty conclusive picture of total human dominion. But this natural shrivelling has not been accompanied by greater strife for humanity – indeed, quite the reverse. Whether measured by health, life expectancy or material wealth, with some notable exceptions we have never had it so good. We seem to have flourished while trashing nature. How can this be so if our lives depend on its health? The answer is that we are living on borrowed time. Many of our natural systems and semi-natural systems like a fresh water supply, hospitable climate and even productive farming are approaching a perilous tipping point. These are ecosystem services that we are failing to appreciate or restore. Like the man who has stepped from the roof, we think, just before hitting the ground: 'So far, so good.'

The problem is that for many people – especially those in the wealthier, more powerful nations – their current lifestyle does feel good. While some understand that we have to change our behaviour to allow more space for the wild things, many people don't get it or want to ignore it, and leaders like appealing to the many. Despite the fact that

scientists and leading politicians know that nature conservation is self-preservation, a little upfront discomfort seems
to prevent actual delivery.

There are non-selfish reasons to look after other life
forms; they simply deserve their own life. I am not a vegan,
a strict Buddhist or even a scratch philosopher but I do feel
that abundant wildlife is a moral good and that we should
follow the 'live and let live' approach. Although there will
be compromises, wherever possible, I believe we should let
the web of life flourish – it was here before us and will, in
some form, survive our demise. When written down, this
sounds a little trite but I suspect these sentiments and this
respect for other life forms are widely shared.

The plight of nature is a global story. Much of what
we celebrate – not least in glossy natural history films –
is overseas and frequently in the last remaining wooded,
arid or ice-bound wildernesses. Some ecologists believe
that natural life in the UK is so degraded that it is barely
worth saving and certainly not if such actions promote harm
elsewhere. Professor Andy Purvis at the Natural History
Museum in London told me that the UK is in an especially
unusual position as we have next to no biodiversity that
'matters globally'. We were one of the first countries to
really get the hang of turning nature into profit through
industrial farming and we soon got rid of any remaining
species that were rare or charismatic. Given such a low
baseline, he thinks we must make sure that conservation
at home does not come at the cost of doing more harm
to biodiversity overseas as exporting damage elsewhere
would be 'absolutely appalling'. Many UK wildlife charities

would say that making space for nature in Britain does not mean shrinking it elsewhere, but I think Professor Purvis's trenchant opinion offers a useful warning against ignoring the unintended consequences of apparently virtuous actions.

Hogging the halo in recent years has been rewilding, which the online Cambridge Dictionary defines as: 'The process of protecting an environment and returning it to its natural state, for example by bringing back wild animals that used to live there. Rewilding runs directly counter to human attempts to control and cultivate nature.' Examples in the UK would be the Knepp Estate in Sussex, Ennerdale in the Lake District or Bunloit beside Loch Ness. Examples elsewhere are Grand Barry in the south-east of France, the Oder river delta on the German–Polish border, and Yellowstone National Park in the USA. Rewilding is, to a greater or lesser extent, the opposite of farming and that is why it is so contentious. In the UK, its chief advocates are the pressure group Rewilding Britain, founded by George Monbiot and Rebecca Wrigley, which now boasts a staff of more than twenty. I went to meet the director Alastair Driver at Broughton Estate – 440 hectares of Yorkshire land in transition – who was wearing a T-shirt with the words 'Born to Rewild'. Beneath the simple slogan lies a much more complex reality, which Alastair feels the need to clarify right from the start.

'Nothing in England and Wales is pure rewilding yet,' he says, 'because we are in a transition. And most of the projects are halfway up the spectrum. Knepp, Ken Hill [in Norfolk], Ennerdale and Hawes Water [in the Lake District] – they're all halfway up the spectrum and none of them will get to the top of the spectrum because we won't have lynx and wolf and elk and boar and beavers everywhere for the foreseeable future. So they're halfway up and therefore they will produce some food.'

One of the things I had assumed was a USP of rewilding – no food production – has just been shredded. I ask Alastair if food production will ever vanish from these nominally rewilded areas.

'If you were to go all the way up to the top of the spectrum and we had native wild herbivores back, then yes,' he replies. 'But that is not going to happen in the next fifty years. And we need to be focusing on, roughly speaking, the next fifty years because we've got this climate emergency and biodiversity crisis. So we really need to crack on and do what we need to do now with the ambition that we will – when the opportunity arises – get further up that spectrum, and that's for future generations.'

'Spectrum' is a word that Alastair uses a lot and that runs contrary to the image of rewilding, which is one of simplicity, almost fundamentalism. Rewilding, it seems, is very pragmatic. Time to clarify with a question: 'When people hear the term rewilding, they think of stopping farming or stopping whatever land management I'm into. I'm going to let nature take its course. You're telling me that's not the starting point?'

'Not in this country,' Alastair says. 'Because in order to rewild – having a balance between carnivores, herbivores, vegetation, soil, water – having a naturalistic balance is not possible because we don't have the native carnivores. So we have to be the wolf and the lynx. We also don't have a lot of the big native herbivores, so we have to use proxies for those. If it was an ideal world and we could magically click our fingers and have these things back in the right places, it would be different. But I'm a realist, a pragmatist, and that ain't going to happen in my lifetime. So we need to look at a slightly less purist approach, which takes us in that direction, delivers multiple public goods en route.'

Rewilding Britain has a network of nearly a thousand landowning members and over seventy case examples on their website. Alastair then says something that, before our conversation, I would have thought was a contradiction akin to virginal pregnancy: 'This is *managed* rewilding. It's working with the tools we've got, which in this case are proxy herbivores – mainly cattle and some sheep – which deliver some food, though not as much. I have calculated the data over fifty projects covering 42,000 hectares, which shows that they are delivering half the meat they used to. It is overwhelmingly less lamb and a little more beef.'

We are about to set out to see rewilding in action and it is already clear that this is a hands-on activity; they haven't just stopped farming and walked away. So, given all this management, how does rewilding differ from other conservation methods? Alastair knows plenty about the range of measures as he was head of conservation for the Environment Agency before joining Rewilding Britain, and he says: 'It's cheaper in the long run because you are not spending lots of money maintaining huge areas of land. And that, for me, is the really big deal. There is not enough money in this country to deliver the scale of nature, recovery and tackling of climate change that we need, if we go down the traditional conservation management route as normal.'

After priming the pump with some hands-on management, you can let nature do the work because it does it for free. So, I ask, does a specific date to let go the reins of control have to be built in from the start?

'That, for me, has to be a fundamental commitment in principle from the landowner or the organisation,' Alastair says. 'When I have this first meeting with them I say, "Look, this is a really important principle. Once we've done the initial intervention, are you prepared to then relax about managing everything, controlling, gardening, in effect, that we are so obsessed with in this country? Are you prepared to relax? You have to be if you are going to continue to move up the rewilding spectrum. If you're not, that's fine. But you're not rewilding, you're either conservation management or nature-friendly farming." That's great – we still need loads of that – but it wouldn't be rewilding. I'm not saying it's a bad thing, it's just different.'

It's time to head up the hill to see what rewilding means here in an estate that was largely grazed by cows and sheep for milk and lamb. Much of it still is, as robust farm tenancies give families strong tenure and some want to stay and carry on farming the way they're used to. We briefly wait behind a truck carrying big tents and metals frames for a forthcoming event billed as 'Europe's Toughest Mudder'. These events across Europe and the US are now a huge business and evidence of lucrative recreational land use. We continue across a ford and skirt a broad wooded ravine to our right that Alastair says is earmarked for a future beaver enclosure. At about 4 hectares, he hopes it will be big enough and the beavers will turn it into a much more diverse habitat. Then we come upon the first example of rewilding in action: tree-planting. Taking land that was pasture and putting trees on it is an act of engagement not abandonment – indeed, very detailed engagement: 'We've got about fifteen species and it wasn't truly random,' Alastair says. 'We planted in zones a sort of high forest species in the middle, lower woodland species around that and then thorny scrub and willow around the edges. So we tried to create a sort of structure that you might find in a more natural woodland landscape.'

A bright green 'tongue' of sheep-grazed pasture intrudes into the middle of the saplings, still in their plastic rabbit guards and growing among scrubby vegetation. They went in less than a year ago. It looks a bit messy beside the pristine sward, I say.

'I'm an ecologist so I obviously don't look at it that way,' Alastair replies. 'It's heaving with invertebrates. You walk into these sites now on a still, sunny day: butterflies everywhere, grasshoppers pinging everywhere, sizzling. The whole landscape in those fields is now sizzling with life and noise and activity. I can tell it's dramatically different. We had an incredible population of Wall butterfly, which is actually a relatively uncommon butterfly – it's one of the Brown family. Here, as soon as we started this, allowing this grass to grow, it just rocketed.'

A little further on, the saplings have been in a little longer, two years, and Alastair is very pleased with their progress. The summer of 2023 in the UK, being quite wet, has been kind to trees unlike the record heat of the year before. Fast growth means the plastic tree tubes can come off earlier and be in better condition for reuse.

We've stopped at the top of the moor to get our bearings. Below is the valley of the River Aire, with Skipton and then Leeds off to the east. Across the vale is a typical British moorland vista: green-brown gently folded hills with blotches of conifer plantation. Towards the valley there are more luminescent fields and occasional broadleaf woodland. Alastair sees current waste and future potential in it.

'This approach that we're taking here could easily be applied in those upland areas because they are predominantly sheep grazing in suboptimal farmland, suboptimal soils. Basic payments are being phased out [the current subsidy regime for farmers], so it's going to be tougher for sheep farmers to make a living just farming sheep. For some of them in the right places, it's probably better to go down the whole rewilding route; others might opt for more nature-friendly farming.'

The ground immediately beneath our feet is tight, short grass, much like a bowling green, and 'mown' regularly by sheep. If any animal is in the crosshairs of the rewilding posse it would be sheep: wanted for crimes against nature and hydrology. George Monbiot coined the term 'sheep-wrecked' to describe what he saw as the subsidised overgrazing and consequent ecological destruction wrought by woolly herds on the hills. But the water problem is more apparent here today: 'We've got film and photos of sheets of water pouring down the slopes above the surface, not soaking in literally because of the pitter-patter of tiny feet for decades. It's meant that the soil is quite compact. But now we've got all these trees in the landscape. You've got this greater porosity. So much more of it is going to soak into the bedrock below.'

The way sheep eat, selecting flowering plants and nibbling very close to the ground, harms biodiversity much more than cattle or horses, so those bigger herbivores are selected for the privileged role of conservation grazing, often afforded the greens' seal of approval. This 'bovine good, ovine bad' policy is reflected in the rewilded land Alastair oversees:

'The entire [livestock] reduction is sheep – down to 5 per cent of what they were – on rewilding sites. And cattle have gone up slightly, by about 30 per cent. So the livestock units overall are down about 50 per cent. And, whereas a third of our lamb is exported, none of this rare-breed cow beef is exported and it's quality stuff, all used locally or regionally. It's not, as you'll often hear, "Oh, you're offshoring food production." No, we're actually onshoring it. We're swapping large numbers of sheep for small numbers of cattle and pigs and rewilding at the same time.'

As Alastair shows me more of their work, it's obvious that huge effort has gone into tree-planting. He says that, in places like this without local copses or hedges to provide tree seeds, stopping grazing would just mean rank rough grass slowly replaced by scrubby ash, hawthorn and sycamore over at least three decades. Most landowners can't wait that long. It's not especially good for wildlife and unappreciated by onlookers. There is another motive, though: some of the money comes from an organisation known as White Rose Forest, which funds woodland creation across the county of Yorkshire.

Yet there is one area where they have let nature take its course: a 6-hectare field closer to the valley bottom that was one of the few arable plots. They took the last crop off it three years ago and it's being rapidly colonised as there are trees growing nearby. Alastair's eyes light up as we push through the dense vegetation.

'It's been amazing to see how it's changed in that short space of time. We knew we were going to get hawthorn, ash and sycamore because those three tree species you see dotted along here. And sure enough, they're all present. But there are also other things. There's a willow here. We've had hazel. I think dog rose has appeared. I think we took seven species of tree and shrub inside two and a half years.'

The sun has greeted us as we wander through this scrubby meadow, and butterflies move ahead of us like a bow wave. Stay still, look and listen, and you will see hovering, humming pollinators reveal themselves. They're all enjoying a much greater variety of flora: 'There's lots of creeping thistle dotted around,' Alastair says. 'You've got rosebay willowherb, which is a non-native but it's a brilliant pollinator for the elephant moth. You've got ragwort, you've got various grasses, you've got buttercups, you've got arable stragglers like fat hen, chickweed and scarlet pimpernel.'

While Alastair has accepted the need for managed rewilding, he clearly loves the places where neither he nor any other human being is taking any decisions other than 'leave it alone'. There is a thrill in the unknown, discovering what life forms have appeared, speculating whether they've always lain beneath. Rewilding can be like botanical archaeology as what was buried takes its place in the sun. Another asset of naturally rewilded sites is the likelihood of survival: the plants that grow are, by definition, those most suited to the spot. Natural selection has decided. But Alastair finds authorities and funders are much less keen to cede decision-making to nature – especially when it comes to trees, where it's all about targets.

'They are obsessed with just sheer numbers of stems in the ground,' he says. 'So this is exactly the conversation I'm having with government. I had a meeting two weeks ago about this. They are concerned that they're not meeting the targets. People aren't taking up the woodland planting scheme like they hoped they would. And it's partly because natural regeneration is not being promoted enough and partly because of paperwork and complexity. But if they were to promote natural regeneration opportunities, they would get trees for free across huge tracts of land. You'd have landowners embracing it because actually they don't need to do much and those trees are more likely to survive because nature has decided: right tree in the right place.'

I get that this is difficult for politicians. In the last general election in the UK there was a bizarre week where the main parties were all competing over how many trees they were going to plant. The Conservatives pledged 30 million a year, the Scottish National Party 60, the Liberal Democrats 60, the Green party 70 and the Labour party a whopping 100 million per year. It was a straight numerical pledge-off. No one said 'we're just gonna leave a few thousand hectares alone and see what happens.' It sounds a bit vague, lazy and out of control.

This target-driven culture for tree-planting has another side effect: millions die. I've seen this in some of my journalistic work. Nearly a million trees were planted beside a new motorway bypass because that was promised, but there was no commitment to *survival*. Planted too close together and without much aftercare, around three quarters of them were dead three years later. Alastair says even conservation funders are corrupted by pervasive number worship. They often insist on 1600 stems per hectare: 'We should not be planting at that density; about half that, 800, would be better. And though, in fairness, they have relaxed a bit on that, the rules still suggest more than they should be planting.' The act of planting a tiny, whippy sapling and slotting on a plastic guard does nothing for climate or nature. Ensuring it grows does plenty. However, most of the money goes to the former.

My guided tour of rewilding in action is coming to an end and my main revelation has been just how hands-on it is. It reminds me of the petite wildflower patches in our garden that require so much attention. It all seems a long way from that definition above: 'Rewilding runs directly counter to human attempts to control and cultivate nature.' But that perception is one of the things that many farmers find toxic. They often describe their work as a calling, not just a job, and rewilding shakes that faith, as Alastair has seen.

'I've worked with farmers all my career. Thousands of them. And some will think it's undermining what they – and their fathers and their fathers before them – did. "You're telling me I've done it wrong all my life?" No, I'm not. What I'm saying is what you did was right for that time. And the policies and the government funding drove you to do that. That might have been right post-War and into the sixties, but now we've got new challenges. If you're up for it, we at Rewilding Britain can help you make that transition by guiding you in the right direction.'

I put it to him that there remains a belief in the 'nobility of farming' and that rewilding feels like a threat to that, both for farmers and our wider society.

'But the important thing to remember is no one's forcing farmers to do this,' he replies. 'We're just giving them the option for their land. It probably means embracing other sources of income, like light-touch nature, tourism like camping, glamping, bed and breakfast, holiday cottages, that kind of thing. But it's not compulsory. It's your choice.'

Considering the divisiveness of the term, and the reality that the land is often still managed for decades in a rewilded space, I ask Alastair if the word rewilding is more of a blessing or a curse.

'Blessing,' he says. 'Recent polls gave us an 81 per cent approval rating for the word and the activity. I accept there are plenty of places where people can't use the word because it distracts people from the proper conversation. But it has an IUCN [International Union for the Conservation of Nature] definition and set of principles. It is a thing and, where appropriate, we should be using the word.'

There is little doubt that the term 'rewilding' implies a rollback of the expanding human civilisation project that has consumed us for millennia. We have overstepped the mark and it's like a telling-off for humanity, so it's hardly surprising that it is controversial. But I think it cuts to the chase and concentrates minds on how we provide more space for nature. Achieving that globally will need much tougher landscape-protection laws and a shift in food culture to alleviate the pressure on wild land, but exclusivity is not the only answer. As we've seen, including nature in our other land-use plans – sharing – works too.

Given that even Alastair is reluctant to prescribe rewilding as the cure on more productive ground, I wonder what routes there are for nature recovery on arable land. Tim Parton, the Staffordshire farmer featured in chapter 4, who is harvesting bumper crops with very low chemical usage, has abundant wildlife too. As I walk with him between fields of vigorous wheat on one side and oilseed rape on the other, the birdsong is constant, which Tim credits to the absence of poisons especially insecticides.

'Many red data [endangered] species are flourishing here,' he tells me, 'like greenfinches and skylarks. I think we had 300 linnets rung last summer. We catch moths for scientists at Rothamsted [the renowned agricultural research centre] as we have numbers equivalent to a woodland. That's because the whole ecosystem is working. With so many moths I have all those caterpillars in the spring. If you look at our bird boxes, the great tits and blue tits will have clutches of eight or ten because there is so much food they know they can feed them. We've reared fourteen kestrels in the last five years.'

I check this with an ornithologist Paul ten Hoeve, who does regular bird surveys and some ringing on the farm. He comes every year with a thermal-imaging camera to count birds at night. Most of them come down to earth in the hours of darkness, and with his kit they show up as little white shapes on the ground as their body warmth betrays them.

'It is like a blizzard of white dots,' Paul says. 'So many and so many species: skylarks, snipe, jacksnipe, woodcock, grey partridge. Animals show up too and we recorded a harvest mouse. "So what?" you say. But that species was said to have vanished from this part of south Staffordshire for the last fifty years. Now the Mammal Society have reintroduced some more with the help of Tim's son, Mackenzie. All this life is thanks to there being so much invertebrate food on or close to the surface.'

In the day he sees reed buntings, yellowhammers, bramblings (a kind of finch migrating to and from Scandinavia), chaffinches and he confirms the kestrel numbers above. Most of them born in specially erected boxes: 'In the breeding season, you can hear the difference on Tim's land too, with calls from skylarks, lapwings and goldfinch. Even farm managers from the RSPB (Royal Society for the Protection of Birds) came out to see what they could learn from his land management. He is showing what can be done. If all farmers took an interest like Tim, we'd all be better off.'

The birds, of course, are part of a wider ecosystem with plants, mammals and insects. Tim has wildflower strips beside some fields and fat hedges containing so many mature trees they look more like linear woodlands.

'All the farm is connected with corridors or hedgerows,' he tells me. 'When I first started not using insecticides, I thought it was vital to have that safe area where my natural predators could be and build up. Now they are just every- where so it's [insecticide usage] not essential anymore. If you walk my crops in the autumn, it is just a mass of cobwebs – your boots would be white with cobwebs. Those spiders are on patrol. If I do get any aphids, they are there to gobble them up.'

This is key to Tim's approach: he wants to share the productive space – the field itself – with the bugs because it is good for his crop. Land sharing here is not a compromise where each side loses a little; instead, both win.

His insect monitor is entomologist Dr Lucy Witter from the Cheshire Wildlife Trust. She walks a variety of set routes on the farm every month and counts bumblebees, solitary bees and hoverflies. Less easily identifiable species are caught and identified back at the lab. Her reaction to Tim's place is gushing: 'In terms of abundance, it is fantastic. And it's not just pollinators. There are plenty of creatures like lacewings, which eat harmful pests like aphids. It's quite unique, and the way he has integrated nature into the farming business is an inspiration to other land managers, especially in this area of the Midlands.'

Surely these high scores from biologists *and* bumper harvests could be replicated across much more of arable farming. The UK-wide 'State of Nature' report, compiled by a number of conservation groups and research scientists, is published every few years and is acknowledged as a fairly accurate assessment of the plight of our natural world. It continues to chart much decline with few recoveries and the prime suspect is identified as intensive farming. If more farmers adopted Tim's techniques, why couldn't the nation have higher yields of both crops and critters?

Tim is a farmer, albeit a nature-friendly one, so I'm interested to ask what he thinks about rewilding, given my perception that many farmers are sceptical at best.

'There isn't the need to do it when I am proving that I can produce food and do exactly that [boost nature] at the same time,' he tells me. 'And I don't get taking land out of production when there are so many people still dying of hunger-related problems. I think it is wrong to take food out of the system because all we're going to end up doing is importing the food from somewhere else and creating another problem in another country. We need to work in balance, to grow food in a way that we are enhancing the natural ecosystem on farm at the same time.'

Alistair Driver might argue that Tim has misunderstood 'rewilding' in the UK context. Maybe so, in which case it seems to me that the term itself is confusing and divisive.

There is one further achievement of Tim's to document on his Brewood Park Farm, though: carbon capture. Everything that grew on the field is left on the field, save what goes to market. That means the stems of the barley, wheat, oats and rapeseed alongside the whole broken cover crops. Most of the carbon captured from the air by photosynthesis is being left on the ground. Of course, some will rot and return to the atmosphere but some will be taken down to be stored. Conventional wisdom has it that, while soil carbon storage has enormous potential, the ground has a limit – a maximum carrying capacity for organic matter. Tim is yet to see it, he says: 'People often ask "Will you get

to the stage when you can't sequester any more carbon?"
And I say no. I've seen worms going down 1.2 metres and it
was just a black motorway where they had taken that carbon
down. The whole organic soil profile goes down deeper and
will just get better and better.'

The journalist in me is searching for the catch. It all seems
too good to be true. Of course, soil type, weather and the
financial state of farms vary hugely, meaning there is not a
simple cookie-cutter approach to applying this everywhere.
But the principles seem sound and the results proven.

In 2021 I originated a series on BBC Radio 4 about
climate change solutions – *39 Ways to Save the Planet*. One
of its heroes was Duncan Farrington, an arable farmer
whose main commercial product was culinary-grade
rapeseed oil but who was proving he could capture more
carbon on his land year on year. It led me to conclude that
climate-friendly commercial farming is quite possible. Tim
has reinforced that belief and added to it: wildlife and food
production needn't be opposed; land can be good for both
at the same time.

I also want to mention a specific and relatively new government policy that demands more awareness and should
welcome increasing amounts of wildlife into our domains:
biodiversity net gain. The British government website says:
'Biodiversity net gain (BNG) is a way to contribute to the
recovery of nature while developing land. It is making sure
the habitat for wildlife is in a better state than it was before
development.' There you have it, right from the get-go: the
assertion that development and nature needn't be enemies.

Such policies are not unique to Britain, as sixty-nine countries have some requirement to make good the damage from development. Indeed, the USA and Germany have had similar legal safeguards in place for forty years. But it is now the law that most new developments in England need to show BNG of at least 10 per cent. In practice that means the wildlife value of a site must be assessed beforehand by measuring its size, condition, distinctiveness and significance in terms of being a local or national priority. Then any harm you do to that site with development must be equalled, plus 10 per cent, in whatever you build. An easy example would be a housing development on an arable field, many of which don't have much wildlife. If you plant hedges, trees and dig a few ponds in your new estate, it will be quite simple to hit or exceed that target. It also acknowledges that old habitat is often more vibrant than new. If you remove 10 hectares of woodland, it is not as simple as replacing with 11 hectares. It could require up to 120 hectares of tree-planting as forests take time to establish. There are further incentives for enhancement of areas that link wildlife islands into more resilient corridors.

The plans have received a guarded welcome from most environmental groups. They acknowledge that development is going to happen and so it is better to reap some benefit from it than not. However, there is some alarm over the permission to deliver some of the gains off-site, meaning you can improve a habitat miles away from your development and it will count towards your percentage gain. That's not much comfort to a local community if their nearby beauty spot is concreted over and the developers get away with tree-planting on a site a couple of counties away. Also, a market has developed in selling biodiversity credits. The government department responsible for planning reckons half of BNG will be delivered off-site, creating a market worth more than £100 million. It was notable that when I put 'BNG units' into a search engine the huge majority of results towards the top were people selling or trading a block of improved nature somewhere. This financial commoditisation of nature feels uncomfortable to me but this may be because markets are not my natural habitat. Yet it definitely raises the question of whether that natural improvement would have happened anyway. Many landowners, before BNG came in, would improve the biodiversity of their land because they loved trees or birds or butterflies or newts. Now they can get paid to do it, will they do any more than necessary or just say 'thank you very much'? Will BNG actually lead to much 'net gain' above what would have happened with business as usual?

The new law only took effect in February 2024 so the proven answers to those questions are yet to arrive. So let's conclude with rewilder Alastair Driver beaming as he showed me his shaggy-looking slope.

'This absolutely sums up why I bring people like you here, Tom,' he told me. 'I brought the minister, the shadow minister here. Look at this transformation. This was a billiard table with sheep two and a half years ago. Now look at it. Rough grassland, trees ten foot high, a wetter and swampier valley bottom. You walk through it and it's just heaving with invertebrates.'

'And when you're just about to turn off the A-road, when you haven't been here for a few months or more, how are you feeling?' I asked him.

'This year was the turning point. When I came about a month ago. I said, "Bloody hell!" This is instant gratification if you're an ecologist like me. To see that with your own eyes, to see barn owls and kestrels everywhere just instantly responding to this bursting of life. Butterflies in some of the thistle patches up here are just astonishing. Last time I was up, there were clouds of them, clouds of butterflies everywhere. It's amazing.'

As I've already said, the best thing we can do for nature is to leave the quarter of the world's land surface we haven't significantly altered well alone, don't farm the wilderness. This is the 'prime directive' (thank you, *RoboCop*). The next thing is to share our dominion better or even return some of it to nature. Rewilding, at least in Alastair's 'spectrum' approach, has a big part to play, so long as it doesn't result in food production losses that are then made up by ploughing up virgin land elsewhere. Avoiding this collateral damage is much more likely if rewilding is only adopted where farming yields are very poor. But nature-friendly farming – especially when practised by farmers, like Tim, who treasure good yield – could help not just our own creatures but also wildlife around the world.

9

SCIENCE

I am very lucky that my broadcasting work allows me to meet interesting and influential people. But, of all the people whose paths I have crossed for work, which one could really change the world? One name stands out above the rest: Giles Oldroyd.

Giles is a modest, sandy-haired giant who looks like he has just stepped out of a 'Soviet Hero of Science' poster. If his research bears fruit and if there was a Nobel Prize for saving the world, he'd be nailed on for one. He is a professor at the University of Cambridge and director of their Soil Science Centre, and his Wikipedia entry says, 'working on beneficial legume symbioses in *Medicago truncatula*', but that doesn't really tell the story. He is creating cereal crops that make fertiliser from the air.

Cereals like wheat, maize, rice, barley and sorghum provide nearly half of the world's food. Farmers' success in increasing their cereal yields is keeping billions of us fed and alive, but it comes at the cost of massive fertiliser use, which is driving climate change, pollution and damaging the soil, as we saw in chapter 4. Many farmers in low-income countries can't afford it anyway, so their yields are often pitiful.

The idea of conjuring fertiliser from thin air may sound fanciful but it is exactly what some plants do already, with a little help from friendly bacteria. The plant group known as legumes – which includes beans, soy, peas, clover, peanuts and lentils – can 'fix' atmospheric nitrogen and make it available for growth. Giles wants to modify the genes of cereals so they can do this too.

Let's take a couple of steps back. Nitrogen is a key component in the cells of most living things – animal or vegetable – and nitrogen gas (N_2) also makes up 79 per cent of the atmosphere. Sadly, as discussed towards the beginning of chapter 4 on arable farming, the chemical bonds in that N_2 molecule are so strong that this form of nitrogen is inert and unavailable to plants, so we invented artificial fertiliser. But plants of the legume family have managed to access usable nitrogen by engaging with bacteria that have the capability to make ammonia (NH_3). To achieve this the legumes perform something of a magic trick by simultaneously providing and restricting oxygen. The key enzyme that delivers nitrogen fixation, nitrogenase, is inhibited by oxygen – it doesn't thrive in normal atmospheric amounts of oxygen – but overall, the process of nitrogen fixation is so energy intensive that the bacteria need oxygen. They cannot be anaerobic.

Giles explains that the legumes solve this problem 'by creating an environment where they tightly control oxygen levels in the nodule, using the same protein that's in our red blood cells, haemoglobin. And what haemoglobin does is buffer up that oxygen, hold the oxygen and release it just at the right concentration that it gets sucked off straight into respiration and is not there to inhibit nitrogenase.'

Cereals already have haemoglobin genes and so his team are working on how to activate these rather than introduce new genetic material. They have also identified the protein, called NIN, that is vital to the nitrogen-fixation stage in the root nodules of legumes and that has become a focus of their engineering efforts. But re-engineering wheat or barley to source nutrition in a new way is a big ask.

I first encountered Giles in 2013 when I was covering the story for *Countryfile* and he had just won a £10 million grant from the Bill & Melinda Gates Foundation. Giles estimates that on-farm use will come around 2030. He thinks that is a bargain in both time and money: 'We've spent £70 million, but that is to try and transform global agriculture and it's a tiny fraction of the sort of investment you see going into medical research. The vision I'm working towards, where you could imagine doubling productivity without inputs, I think this is all possible, is all in our reach. I'm confident we can deliver this.'

Successful delivery for Giles is not primarily about eliminating fertiliser use in Europe and the US, it's about helping poor farmers in Africa. A shortage of nitrogen severely limits smallholders' yields. Fertiliser comes first from the farmers' own waste; some may have access to livestock manure and very few can afford artificial fertiliser. Average nitrogen application in sub-Saharan Africa is 3–5 kg per hectare, whereas the world average application rate is 134 kg per hectare. A relatively modest success in getting cereals to do their own nitrogen fixation could easily double harvests and have a huge impact. That is why the Bill & Melinda Gates Foundation is involved: they see boosting agricultural productivity as a key way to achieve their goals as an equity and poverty-reduction charity.

Such success could have profound environmental benefits too. Africa is where population growth continues to be stubbornly high: the continent is expected to add more than 1 billion people to the world's total by 2050. Reaping more grain from every farm would help to feed those extra mouths while keeping wild areas wild. Food security and wildlife protection without worsening climate change – see what I mean about a Nobel Prize?

To put an even firmer brake on climate change, though, you need to slash fertiliser use in the world's breadbaskets and, if Giles's project is successful, this is well within reach. The amount of nitrogen that legumes get from symbiosis with mycorrhizal fungi is more than sufficient to deliver all a cereal crop's needs, he says: 'In fact, cereals need less nitrogen than legumes, as they have much less protein in the seed. Hence, if we are able to transfer this association [nitrogen fixing] in full to cereals, then we would not need nitrate fertilisers. I suspect we will not instantly get full levels of nitrogen fixation; it's more likely we can increase this over time as we increase efficiencies, hence I expect a stepwise replacement of N-fertilisers.'

It would be an achievement on a world-juddering scale, upending many of the foundations of farming and totally remoulding the relationship between food growth and nature. It would mean more food from less land while shrinking the collateral damage of the climate and wildlife. Giles is confident, and I passionately hope he's right, but self-feeding staple crops can't be regarded as a sure thing. Giles sees his own particular task as part of a larger effort within crop science to work *with* our environmental goals, not against them. The key is replacing chemistry with biology.

'If I look back at the last hundred years of agriculture,' he says, 'breeding has contributed a lot but chemistry has contributed even more. And it's just because chemistry was available and an easy way to deliver those solutions. Actually, if you look in the natural world, plants have solved every problem we have in agriculture. It's just not always in our crop plants.'

His approach is to discover the genetic basis of some plants' abilities to deliver solutions like drought tolerance, disease resistance or faster growth and introduce them into our food crops. If that means genetic engineering, and tests show that the particular application is safe, then so be it. But he knows that at the moment the private sector has little incentive to make this shift to more biological solutions, even those not involving genetic engineering, as they are doing well from the status quo: 'Industry is happy where things stand, because most of your big biotech firms are producing the fertilisers, producing the agrochemicals and producing the seed. So they're not in the game to change their markets around radically.'

One of the most environmentally damaging links is between crop varieties and chemical usage – a toxic relationship that needs to end. The Cambridge Crop Science Centre is trying to develop a more sustainable relationship for crops – this time with fungus. As we've seen, nutrient-swapping between plants and fungal allies is the basis of floral life on

Earth. But chemical fertiliser gives the crops food for free so the deal is off: the plants thrive and the fungus shrivels. In effect, the fertiliser starves and therefore kills the fungus, its coffin nailed tighter by regular doses of fungicide and ploughing. Commercial arable fields are almost unique among terrestrial ecosystems in that plants can flourish in the near-total absence of fungi. (Incidentally, bogs are another one because mycorrhizal fungi can't survive being waterlogged, so many of the aquatic plants turn carnivorous and get their nitrogen and phosphorus from devouring insects.) On typical cropland you'll see short, shallow roots that have no need to reach down to find food or talk to mushrooms about a nitrogen swap because the field surface is being regularly showered with easily accessible nutrition. Giles says plant scientists have helped cement this crop co-dependence with chemicals: 'Mycorrhizal fungi have been totally ignored [by plant breeders] because all the breeding is done under these high-input conditions where the fungus isn't even present. So there's no benefit; if anything, they would be breeding against the fungus.'

Crops don't just become hooked on the chemical fix, they're bred for it. It's actually in their DNA. And yet, according to Giles, the complicity of agrochemical companies, science and government goes further.

'All the commercial companies put new wheat varieties on the market each year,' he tells me. 'NIAB [the National Institute of Agricultural Botany] test them all side by side and put their recommended list together, showing which lines performed better. But it's all under high-nutrient conditions. Even DEFRA [the UK government's Department of

Environment, Food and Rural Affairs], which talks about
sustainability, is paying NIAB to test wheat varieties only
under these high-nutrient conditions. So the whole system
is stuck in a way of farming. That is the current way of
farming. But it just happens to be the one that was invented
because chemistry was ready to solve the problem.'

Imagine the celebrated TV show *The Wire* relocated
from the drug cursed streets of Baltimore to the global
farm. Plants once worked for a living, trading carbon
derived from photosynthesis with fungi, which paid for
it with nutrients like nitrate. There was work for all and
a thriving biological community. Until dealers moved in
offering a chemical fix – nutrients for free and you can
get higher yields. So, with no market for what the fungi
were 'selling', they vanish. Then chemical suppliers started
breeding customers to be totally reliant on their product.
Those same companies also sold them chemicals to deal
with the illnesses worsened by their narrow chemical diet.
If they tried to go cold turkey, and grow without synthetic
chemicals in their veins, they'd be yellow and stunted.
Meanwhile, the authorities only recommend to farmers
those varieties which have been bred for chemical addic-
tion. Such is the dystopian truth of most farming today.
But now we must pay the full price of those chemicals and
we've realised they'll cost the earth.

I wanted to check this story with Dr Phil Howell, who
is head of breeding at NIAB and also spent years working
in the commercial sector. He confirms that the testing
system for new wheat varieties is run with pesticide use
above general farm practice and crops are 'soaked with

nitrogen *above* farm optimum.' I am staggered by this. Everything, even the official testing regime, is stacked in favour of chemically addicted strains and against those who can thrive in a healthy soil. Phil doesn't quite share my increasingly high-pitched outrage, but he wholeheartedly agrees that the current system hinders the breeding of more environmentally friendly strains of staple crops.

'Until we see wholesale changes to the testing regime, that is the biggest stumbling block,' he says. 'Voices are beginning to be raised but there are lots of vested interests and the entire supply chain needs to recognise the need for change before we get change.'

With all the buzz around regenerative farming, I ask why we are not seeing crops being developed for that lower-chemical system. Phil replies: 'As a former commercial breeder, I know they are breeding for demand, for volume, and – despite the noise – regen farming is still quite niche. The royalties you make from farmers using your variety are quite small. Niches aren't commercial.'

This strikes me as the most obvious candidate for state funding. Alternatively, we should ask the big beasts of food supply, who are beginning to claim regenerative credentials, to put their money where their mouth is and pay for the development of varieties that are suited to that system. Oh, and we should overthrow the testing regime.

For many people, the alternative nirvana is organic farming, which shuns all of those chemical inputs, yet we know that you can't feed the world's population (let alone a growing one) on organic farming. Giles is as caustic about the idealism of organics as he is about the perils of Big Ag.

'Organic is less than 5 per cent of global productivity, despite a lot of noise for many decades,' he says. 'And they act as if they own sustainability [and] behave as if you can only be sustainable if you're organic and you follow all these rules. It's really unfortunate because 95 per cent of the planet is not organic and that's where the environmental impacts come from. I want to see more sustainability where it really matters, which is the 95 per cent of productivity. And organics have barely touched cereal production. Organic impacts the small, high-value commodity crops: coffee and chocolate and fruits.'

Giles and many others see regenerative agriculture as a philosophy of farming that respects both nature and productivity with science as an ally: 'I'm super-excited about regen agriculture and I've been talking to the regen farmers, trying to find ways to work with them and devise projects to bring genetics into this farming process.' Adding that one of their core goals aligns perfectly with his, he says, 'I think, if you consider what is a sustainable farming system, it has to be less fertiliser across the board.'

Regenerative farming tries to rely much more on natural soil biology and fungal alliances for crop growth, but what variety of wheat is suitable for that? As the commercial types are optimised for maximum chemical dosing, they won't be inclined to grow deep root structures or make deals with mycorrhizal fungi. The alternatives are so-called heritage varieties, which – as the name suggests – are little changed since ancient times. A good example is spelt, a species of wheat that would have been familiar to our Bronze Age ancestors. These varieties' historic cachet and claimed health benefits mean they can command a decent price, but the productivity is, in Giles's words, 'pathetic' – useful for feeding the affluent British shopper but not the world. That puts the responsibility on plant breeders to bridge the gap with modern strains that are not hooked on chemicals.

'What we need to do is design the varieties for that type of farming,' says Giles. 'And the lines that we're putting out in the field right now are perfect. They farm for mycorrhizal fungi. I've got varieties that maximise that engagement.'

The field in question is just south of Cambridge, near the village of Duxford, also home to the airfield base of the Imperial War Museum's collection of aircraft. In 2022, the team from the Crop Science Centre and NIAB planted a type of barley genetically modified to improve its inter- action with soil fungi. Specifically, it has been modified to boost the expression of the NSP2 gene, which should enhance the crop's existing capacity to engage with mycor- rhizal fungi. They're also planting control varieties that are unaltered and another that has the fungus-friendly gene suppressed. They hope their principal GM strain has the potential to replace or greatly reduce the need for inor- ganic fertilisers, with significant benefits for improving soil health while contributing to more sustainable food production. Results should be in by the mid-2020s. It is all very promising; however, Giles spies another worrying trend in land-use policy.

'If you're going to say that 30 per cent of the land needs to be managed for biodiversity,' he says, referring to the commitment made by the UK government and nearly 200 others globally as part of the Global Biodiversity Framework, 'then what's remaining for food production needs to produce food. Otherwise, all we do in Europe is ship demand to Brazil. We end up with more skylarks here and a thousand fewer species in the Amazon because we cut the forest down to compensate for the under-production. If in Europe we just change our landscape to something that is supporting wildlife, we get productivity from countries that don't care. Good intentions, bad result.'

He thinks specific areas should be better managed almost exclusively for wildlife – like national parks – but where we farm, in order to avoid pushing environmental damage elsewhere, we must reap a lot of food. He believes that sparing land for nature, rather than sharing it with farming, is the optimal thing for biodiversity and crop yields but only if the wildlife is properly protected in its realm and food is prioritised on the farm.

Yet this movement against the grain of conventional environmental attitudes feels like a smooth ride compared to the reception he got from those who oppose genetic engineering. In the firm belief that he was doing exciting yet safe science that would help poorer farmers and slow climate change, Giles was initially happy to publicly argue for the potential benefits of GM, but the backlash has made him step back.

'I find it so frustrating that for the entire time of my career, the guys who are promoting sustainability won't talk to me,' he says. 'They think I'm the evil guy because I'm going to use some new technologies in crops in order to achieve sustainability. I'm not part of the discussion as far as they're concerned. And it's bullshit that I'm going out to help Big Ag. I'm not working for Big Ag, I'm working for a charitable foundation whose intent is to raise productivity for smallholder farmers. When we spoke ten years ago, I was doing a lot in the media and I was just fed up of being the bad guy against the Soil Association or Friends of the Earth. Actually, we're all driving for sustainability and I got bored of being the "evil scientist" when what I'm really trying to do is drive sustainability and I think I'm being realistic about it.'

What's on the floor of Dr Lydia Smith's office reveals more than what is on her desk. She's allowed a little private 'hutch' that we reach across the open-plan sweep of desks pleasingly colonised by indoor plants (which gets a big tick from me for combining workspace with nature space).

'That is an ex-spade,' she tells me, pointing at the rectangular steel blade now lacking a handle. She had been trying to show a group of Australian farmers what lies under the surface of a research plot growing multi-species grassland. It had been very dry: 'I managed to dig up a big sod of earth to have a look at the root structure at the one site. And then we came to the second site and I tried to do it again and I literally broke this in half.'

Next to the spade with its jagged, severed neck is a plant that's illegal for farmers to grow in the UK: hemp for cannabis oil. It's about a metre tall and looking a little wilted after a few days out of the earth but still has the trademark fan of saw-tooth leaves – the beloved counterculture badge. When lifted up, a fine fog fills the space between us. Not a psychoactive illusion but simply pollen:

'It's an absolute record breaker. It produces more pollen than practically anything else.'

In fact, there is nothing in this hemp to make us high as it has virtually none of the mind-bending compound tetrahydrocannabinol (THC), just cannabidiol oil (CBD), the active ingredient in delivering some pain relief. THC is a recreational drug with known side effects, whereas CBD is an increasingly popular medicine – a distinction still apparently lost on the UK government, which outlaws both: the only hemp types permitted to take root are low in both the ingredients above and useful only for fibre.

Lydia likes to get her hands dirty and challenge convention, and she has just been chosen to lead a research team 'set to increase carbon capture through cropping', according to their official press release. It's a multimillion-pound project part-funded by the government to create the Centre for High Carbon Capture Cropping, going by the somewhat tortuous acronym CHCx3. The release states that: 'Farmers and associated industries can address climate change goals through input-efficient crops that are able to increase carbon capture, but they must have confidence in achieving profitable and sustainable outcomes'.

Using the latest science to combine carbon capture with productive farming is my multifunctional bag and it brings me to Lydia's lair beside her research farm on the outskirts of Cambridge, where she tells me: 'We are looking to see how you can pull as much carbon back into the soil and into crops. And then when you've got those crops, what can you do with them that holds that? We're looking at a set of four types of crops that we think could be quite big hitters.' Those four are: fuel crops like willow, poplar or miscanthus (elephant grass); cover crops that share the field with the food; perennial food and feed; and fibre crops that give the field a rest from intensive food growing but still yield an income.

Plants are CO_2 sponges but, for a farming system to not just capture but actually *store* carbon, *it must either build up in the soil or be held in a product made from the crop*. Lydia takes me to a trial plot of a fibre crop that can both deposit carbon underground and leave its 'body' to good use. No prizes for guessing: hemp. One of her many roles is as head of NIAB Innovation Farm, an area just outside Cambridge, eyed hungrily by builders but currently the preserve of high-tech greenhouses and blocks of different crops being grown in field conditions. The hemp stands are each roughly the size of a tennis court with the plant standing about shoulder height, not yet fully grown. Lydia sees the fact that they are beginning to test different varieties, paid for by the government, as a sign that the authorities are softening their opposition: 'I think this is a bit of an indicator because DEFRA sees this as a potentially up-and-coming crop again, which does need to be tested in terms of the quality and usability of varieties for English farmers.'

Hemp has at least three potential commercial uses from different parts of the plant (and that's before you get to the leaves in the more psychoactive varieties): the seeds can be used for medicinal oil, the core of the stem for insulation and the outer husk for fabric.

The first records of hemp in clothing date from China thousands of years ago. Its use then spread along the Silk Road and was commonplace in Europe by the first millennium of the modern era. As European empires emerged based on naval power, the strength and salt resistance of hemp made it a key ingredient of rope and sails. Queen Elizabeth I decreed that, for every 60 acres of land, farmers must grow at least 1 acre of hemp. A number of sources suggest that the English county name of Hampshire comes from hemp (*hamp* or *hampa* being the translation in many Nordic or Germanic languages). The British Empire was held together by hemp. Its decline came from the advent of cotton that was easier to work for clothing, petrochemical-based plastics and nylons for ropes and sails and the exploitation of some varieties with high THC load as a recreational drug (the latter leading to black-market commercial opportunities but also widespread bans).

However, hemp's physical properties are being admired once again. Lydia challenges me to break some hemp strands thinner than spaghetti. I fail the task but not before leaving red ruts in my finger joints. Types of fibre board and plywood are using it and there is a small but vibrant clothing sector. It used to be 'tough but rough' – a bit abrasive on the skin without very laborious working – but now mechanical processes can do the softening job. At the same time, its great rival cotton has been shown to have a punishing environmental impact with high chemical and water usage.

'One of the things that we're doing in our project,' Lydia tells me, 'is working with a company called Fine English Cottons based up in Lancashire. They've taken over one of the old Lancashire mills. The reason that the company's joined us is that they want to see if they can reduce the amount of cotton – which has got quite a few non-sustainable attributes, shall we say – by using hemp. It's like a tougher form of linen… Essentially, you bash it up and you make it softer and fluff the fibres, and then it's much more like cotton. I've got a hemp bag and a hemp shirt I've had for longer than my daughter's age. She's twenty-five and I wear it a bit every year and I wear it almost exclusively in Scotland because it's good for keeping the mozzies and midges at bay without making me too hot. So I would say it's a minimum of thirty years old. And all that's happened is I've just lost a little tiny bit of the cuffs.'

It's not a long-term carbon-storage solution, but a few decades is still useful and, probably more importantly, it adds to the commercial viability of hemp to help pay for some of the below-ground carbon storage.

Then there's the 'shiv': not an improvised bladed weapon favoured by prison villains but the core of the hemp stem. Once the artery for conveying water up the living stalk, when dried it's like a stiff sponge, very light and with myriad air pockets.

'The shiv can be used in lots of building outcomes. Hempcrete is the classic one that's been developed over the last few decades. You break it up into slightly smaller pieces and coat it in lime. And you'll come up with a block that looks a little bit like a breeze block, but lighter and absolutely full of shiv. It's an incredible insulator. If you build a building out of hempcrete you have an automatically insulated building, and some carbon capture and that lasts for as long as your building lasts.'

It's also processed into animal bedding in place of straw, made into paper pulp and rolls or panels of thick insulation for walls or lofts. This is definitely commercial: type 'hemp shiv' into a search engine and most of the results are items for sale. However, the really booming market is for the CBD oil made from the seeds, available as a skin cream or edible oil, which is promoted for conditions like anxiety, high blood pressure and insomnia but especially for relieving the pain of arthritis or myalgia. The marketplace for CBD in the US alone is currently valued at more than $5 billion and expected to grow at least fivefold by 2030; similar growth rates are forecast for Europe. That's going to need a lot of hemp. In the European Union it is now legal to grow hemp for oil, but not in the UK (where it is perfectly legal to sell or consume it). That appears like a straightforward handicap to British farmers. Within the EU, hemp cultivation expanded by 60 per cent between 2015 and 2022 and yield went up by 85 per cent.

So it's clear that hemp is useful, but how does it fit in to the climate and land-use story? The big deal is below ground. Hemp has a thick, tough, tap root and a dense network of finer roots. When the plant is harvested this underground biomass stays underground. Some will be lost as it rots but what remains will steadily boost soil carbon storage. Lydia digs up a fairly stunted specimen (without breaking the spade), about half a metre tall, and even that has a lot to be proud of down below: 'You can see the really, really solid, thick, robust main tap roots, but then all these little fibrous roots. If you want to know the potential for the carbon capture of different crops, it's pointless just looking above the ground. You really need to know what's going on below the ground.'

Root mass below ground matters to the planet in terms of carbon storage but it also delivers more immediately for the farmer in two ways: firstly, all that decaying organic matter feeds soil life; secondly, that gnarly tap root is like a vegetable drill bit boring down into the earth and opening up compacted soil. Compacted soil has been squished together by the weight of repeated traffic and creates an almost impenetrable layer where plant roots, fungal strands and even water struggles to get through. The result is poor yield, more flooding and more drought. Hemp roots can bore down into that solid soil and return the essential gaps. Lydia shows me the root again: 'That's giving you all that porosity that we hope for. And this is a very bad field on our farm which has been overworked, with too many trials over too many years. And that's doing an absolutely brilliant job of breaking up all those compacted bits.'

Lydia and the team at NIAB think farmers should be encouraged to use hemp as a break crop. After a few years of growing the rotation of wheat/rape/barley, they could then plant hemp for one or two years. Along with all the soil advantages detailed above, its speedy growth and dense canopy also helps to eliminate black grass – a weed scourge across much of eastern England. But the scientists know farmers are reluctant to take any land out of production and away from income generation. However, hemp can help here too, Lydia says:

'If you've got a field that has been used primarily for arable combinable crops, i.e. wheat, barley, rape, and you've got to do some years of fallow because the soil quality has just bottomed out, surely it's better to have an economically viable crop that you can get something for and is actively improving it [the soil].'

She would like to see the area of hemp increase tenfold – from 8000 hectares today up to 80,000 – and knows that would be helped by a shift in public perception and government policy: 'That would mean you would start to see it all over the country and it would be like a snowball. Once it becomes accepted, and people aren't frightened and screaming in the ditches that they've seen a cannabis plant, then it adds impetus to the government to be a little bit more relaxed about the licence.'

A few hundred metres away from the hemp plots is another tall patch of green but this one – far from having an infamous appearance – is completely new to me and almost totally novel in Britain. It's a cereal that could be a very big deal because it's perennial. Nearly all our staple food crops are annual, meaning that they live for less than a year and then are replanted: sow, grow, reap, repeat. Wheat, potatoes, maize, rice, peas and beans – they never make their first birthday. We do reap some food from plants that live longer than a year, but these tend to be soft fruits like raspberries and blackcurrants or tree crops like apples or palm oil. It's another thing that makes our arable landscape so unnatural as it alternates between bare soil and youthful plants. A switch from lifeless to infantile and back again. Maturity is unwanted; the first flush is the richest.

Yet the climate impact of perennial crops is far preferable: they require little or no fertiliser, much less polluting tractor traffic and little soil turnover with its consequent carbon losses. The respected climate-change solutions resource Project Drawdown reckons the expansion of perennial crops could pull hundreds of millions of tonnes of carbon from the atmosphere every year. But these are mainly tropical staples like bananas, breadfruit or avocados. What about a cut-and-come-again cereal? A wheat bush, if you like. That's what I'm walking through with Lydia, up to my neck in it:

'This is Kernza, a perennial wheat grass. You keep cropping it repeatedly for, we think, at least six years. And then all of that beautiful root structure and ecosystem services that you get from a perennial crop – which you don't for an annual crop that you plough up and start again – you can potentially retain that. If we can look at some alternative food sources that don't require the very heavy tillage... I think that's quite exciting.'

Kernza is actually a trade name from the American Land Institute. Its scientific title is *Thinopyrum intermedium* and it's not that closely related to wheat but its seed head is similar, hence being known as perennial wheat. The team at NIAB imported some strains from the US and have been growing it for just two years – time enough for harvesting and baking: 'It contains gluten, so you can make bread. We have made bread and it's very tasty. It has a bit of a cinnamon kind of a flavour.'

We are walking through a tall, swaying field of stems, the thickness of spaghetti, topped off with a grain-bearing strip that is about 20 cm long but rather sparse. Its longer and thinner than wheat, though we are at least a month away from harvest. The trial plot is about 8 metres wide but more than 100 metres long. It is divided up into blocks of differing varieties to see which performs better, as it is usual for crop varieties fresh from the US to struggle on this side of the Atlantic.

So far Lydia and her colleague Phil Howell, who leads on this project, are very happy. The second-year growth is better than the first, disease appears non-existent and weeds are scarce. But with yields running at about one tenth that of wheat, it ain't gonna fill your breadbasket – yet. It has taken millennia to turn wheat from a spindly grass found in the Middle East to the agricultural cornu-copia we know today. Lydia and her team are hoping their breeding skill will allow some rapid improvements to be made with Kernza:

'It's never going to give us the sort of 10–12 tonnes per hectare yield that we can get from our highly improved wheat varieties. But you've got to start somewhere. And that wheat tonnage only comes with all of the energy and the human-hours and the chemicals that you have to put in every single year of an annual crop. If you can have a perennial crop that could be in for six years, that's got to be worth considering.'

In the meantime, Kernza is already delivering the similar kind of soil improvements seen with hemp, and Lydia can envisage farmers using it in marginal fields that need help below ground. Her spade is at hand and soon deployed:

'These roots are much thicker, much more robust than wheat roots, and you can see I've in no way got to the bottom. I snapped off nearly all of these ones here. They're

going a lot deeper. But there's a huge quantity of root structure right through the first 20–25 centimetres. All of those
things I say you desperately need in a good soil structure.
Look at the porosity, all these little holes.'

Barring a revolution in arable agriculture, perennial
plants are likely to remain a minority player: certainly worth
encouraging and expanding but far from the dominant land
use. Across the world and in most of the big farming countries it's pasture for grazing, not crops for harvest, that takes
the lion's share of the earth's agricultural surface – globally
around two thirds. Most of these grasslands are already
swathed in perennial plants but Lydia and her NIAB team
believe these lands could be doing a lot more to store carbon
by planting new varieties, as she tells me: 'A high carbon
capture mixture of forage species that will give you a much
higher biomass per unit area of ground than a very simple
mixture like grass or clover.'

Giving grassland a greater appetite to capture and hold
CO_2 would shrink farming's carbon footprint dramatically
and, as many livestock farmers would be keen to point out,
help to shift them from the climate naughty step where
they've been languishing sulkily since the scale of their
herd's methane emissions became known.

'This is something we've been working on for some years
now,' Lydia says. 'If you've got a huge range of different
roots and a huge range of different canopy types, you actually have got a lot of different niches that different plants
can take and make use of. Whereas, if you've got all the
same type of root structure, which goes to all the same
depth, the biomass potential is less. So there's a whole

range of ecological benefits as well as the very simple biomass production.'

These multi-species grasslands are often known as herbal leys, especially where they are deployed in arable systems to give the soil a break from constant tillage. We saw this with Rachael and Geraint Madeley Davies in chapter 5, but how much more could science add to the job already being done by smart stockmen and women?

'We are analysing performance of the different plants within a ley so farmers can have greater confidence over what would work for them,' Lydia tells me. 'These multi-species seed mixes are expensive and maybe only half of them are of any use to you. Greater use will come from greater certainty.'

But while much of this comes down to spreading best practice and shifting from a ryegrass monoculture to greater variety in the sward, Lydia has a radical ace up her sleeve – and it's one that is entirely new to me: sainfoin.

'Sainfoin is an exemplar of a lot of the things that we should be doing in UK agriculture,' she says. 'I think that's why I'm almost neurotic about it. It encompasses a lot of the things that are sustainable and yet productive. And when I say sustainable, I don't mean "tree hugging" sustainable, I mean sustainable in terms of an economic outcome for farmers as well as the environment being maintained in a good state.'

Sainfoin looks a bit like a lupin, with a pink flower that ranges from dusky to puce. It is a legume so it can fix its own nitrogen for fertility from the air, has a high biomass and it's very deep-rooted, giving it exceptional drought tolerance.

Though it originates from central Europe, it was very widely grown across the continent especially in northern France. Its name comes from the French and means 'healthy hay' as it helps the wellbeing of ruminants in at least three ways. Firstly, it makes sheep less likely to explode. Let me wind back… The other popular legume in pasture is clover, especially red clover, with its high biomass and protein content. Overindulging in clover can trigger an excessive growth of a micro-organism that lives in the rumen of cows and sheep and produces a foam that can blow up the unfortunate animal's belly. Fans of Thomas Hardy's *Far from the Madding Crowd* might remember Gabriel Oak lancing the drum-taut flanks of Bathsheba Everdene's flock as they lie dying. His cool head and farming knowledge save the sheep and his brooding strength wins her heart. That was bloat, which sainfoin avoids, says Lydia: 'You'd never get that with sainfoin because the tannins – the complex, high molecular weight tannins that are a natural part of the foliage of the sainfoin crop – prevent that micro-organism overgrowth. They mediate the growth of the bacteria in the rumen.'

This same process also makes the protein in the organic matter more available to the animal so less of it is turned into ammonia. That means faster growth for the animal, less methane emission and lower nitrate pollution from the excrement: 'Although sainfoin hasn't got an absolute bucket full of protein itself, it should be looked at differently because what it has tends to be useful to the animal.'

But the medical properties of sainfoin are not done: a third benefit is that it is an anthelmintic, or a wormer, Lydia tells me:

'Those tannins also stick themselves to the coat of the parasitic nematode worms that nearly all ruminants will get. And if they're all sitting on the same field, as tends to be the way now with permanent pasture, they just eat the eggs from the worms, reinfect themselves and it's quite life-threatening. If you see a sheep, especially a lamb, that's been on infected land and you see its bottom completely covered in poo, that's usually a sign that it's got quite a bad parasitic load.'

The worms themselves can burrow into the folds of the rumen and suck blood from the animal, making it dangerously anaemic. Farmers regularly use wormer but it both undermines the animals' natural worm defence and, as some of the chemical wormer is excreted, it can also kill earthworms and dung beetles. The tannins in the sainfoin don't totally eliminate the gut parasites but reduce them drastically. Lydia tells me how:

'Tannins stick to the outside of the nematode worm. They stop it from going through part of its life cycle. It can't get out of its coat, it can't shed its coat, so it can't reach sexual maturity. Also, tannins stick to the mouthparts of the animal. You've got the biters that bite bits off the rumen, and then you've got the suckers that burrow their way into the folds of the rumen and suck the blood from the animal. It damages their mouthparts so they [the worms] can do neither of those things so well. And that's really important because, if you purge an animal of all of its worms and it's completely worm-free [like the chemicals do], then it loses its natural capability to deal with a parasite load. Whereas with sainfoin it just pushes it right down to a very low level.'

With a healthier animal, more soil carbon, better soil structure and less poisoning of soil life, sainfoin's balance sheet is looking strong. But there is more: insects love it. The Soil Association website says: 'Its extended flowering period and the fact that sainfoin produces more honey than any other legume makes it invaluable to pollinators, attracting a wide range of bumble and honey bees, butterflies and many other invertebrates. Bees feeding on sainfoin produce higher yields of honey.'

Lydia says it was once very widely grown in East Anglia and chalkier parts of Britain because its deep root and nitrogen-fixing capability meant it was such an effective soil improver that landlords would often write in the contracts of farmers that they had to grow some sainfoin. It would maintain the farm's soil quality for future crops. But this motive was lost with the advent of cheap chemical fertiliser and so was the sainfoin. However, Lydia's tireless promotion of sainfoin has worked. In the last twenty years or so it has gone from just being in a handful of fields to growing in leys across much of the UK and becoming a standard ingredient in many seed mixes. And she now even has royal patronage as it was mentioned favourably in a speech from Princess Anne in a recent visit to NIAB to promote under-utilised crops.

Cover cropping is a well-known method of holding more carbon in the soil, but it has much more to offer. Such crops are increasingly being seen as not just regenerative farming practice but sensible farming practice. Lydia Smith and her team are focused on making them fit with changing farming practices, so that there is a greater variety of crops and less chemical usage: 'We want more different main crops but this needs to dovetail with a greater range of cover crops. The different growing season of each must fit together. We are looking at plantain, because it is nutritious forage, is cold- and drought-tolerant, grows well in much of the UK and has a dense root system.'

One of the challenges with cover crops is how to kill them in advance of planting your money-maker. With ploughing and poisoning both having big drawbacks, what's left? Crushing: squishing the plant with a roller. Tim Parton, who we met in chapter 4, was engineering a type of roller to do just that, and now Lydia and her team are trialling different cover crops to see which are easily killed off by simply having the main stem snapped. Mechanical engineering and plant science could combine to deliver a climate- and wildlife-friendly outcome.

Crops grown specifically for fuel fall under the CHCx3. Given my scepticism about biofuels, I can't resist asking Lydia if she thinks they really are a good use of land when solar provides you with so much more energy per hectare. Her response focuses on how good willow can be at improving the soil structure and its potential for building materials. When pushed on the energy angle, she reminds me that the sun doesn't shine all the time and biomass is one of the ways to get away from oil- and gas-based heating. But I don't get the feeling that she's a cheerleader for biomass energy at scale.

Her team reckon those four land uses – fibre crops, perennials, cover crops and fuel crops – are the four legs of a growing carbon-capture platform. Their research is not just searching for or engineering better varieties but also discovering economic blockages that might be arresting their development. 'Improving the value chain', as business folk like to say, meaning making more money.

'For each of the crops we looked at what would success look like,' she says. 'And that would be increasing the amount of each of those four crops types by a percentage. So we felt for the hemp, we were looking at an absolute minimum of a ten times order of magnitude [growth] and for energy crops at least four or five times.'

For cover crops and perennials, there is not so much a growth target as an aim to discover and disseminate best practice in order to avoid some of the bad experiences and give farmers the confidence to move away from unsustainable behaviours that have become the norm in recent decades. So what could this all add up to if Lydia's recipe of farming for carbon storage was widely adopted?

'We actually calculated that we could increase the carbon capture by 65 million tonnes per annum if we pushed up to the levels that I suggested,' she says.

This is a very big figure, more than the total emissions from UK agriculture, which sit at around 50 million tonnes of CO_2e. (The 'e' refers to carbon dioxide 'equivalence' as most of agriculture's climate impact actually comes from methane and nitrous oxide emissions.) It would make farming in the UK a carbon sink. But that nagging question remains: will these practices suppress food production here so that we import more and cause more greenhouse gases to be emitted somewhere else in the world? Lydia thinks not.

The proposed land take for hemp and flax would be around 10,000 hectares: less than 0.1 per cent of UK agricultural area. Willow and miscanthus might grow somewhat but will largely be confined to poor-quality land. It is difficult to put a measure on increased land use for the spread of herbal leys as some will be in existing grassland and others will become part of an arable rotation, but there is no denying it will take some acreage away from growing food. The counter argument is that many of those fields have become exhausted with rapidly declining yield so they can't go on as they are. Putting in a break of a year or two or three with hemp or sainfoin would enable them to recover and produce *more* food in the future.

'The vision is how do you integrate them [the four remedies Lydia favours] into a food production system. Foreseeably, we're still going to need to an awful lot of wheat of the regular type and potatoes and barley and so forth. But it's foolish to think that you can just grow wheat, followed by wheat, followed by oilseed rape, possibly followed by beans, followed by wheat, and expect the soil to be able to carry on functioning as a proper system.'

Synthetic chemicals have enabled and supported highly productive yet simple farming. But its end is nigh and the future is about variety.

'It's really important to push diversity,' Lydia says. 'As opposed to just biodiversity, we also need crop diversity. We've got to have lots more different crops, much in the way that farms used to be managed. And animal husbandry is a really important part of that because, even if you're not growing an animal to eat it or take its milk, you get that interaction of vegetarian animals putting their faeces back onto the land and giving an injection of complex nutrients. But now we have the situation of only arable agriculture in the east of England and very heavily livestock in the west. And we're not benefiting from the different components of a mixed system. But it's not just the presence of the animals themselves, it's the presence of their forage and these complex leys in which there are many flowering components – then you automatically have better resources for pollinators. If you are expecting your oilseed rape crop to be properly pollinated, you need pollinators available.'

The livestock help the crops, the crops help the livestock and the combination helps boost wildlife and cut pollution.

Much of this chapter has been based on the fact that artificial fertiliser is the villain of the piece, the habit we must kick. But what if we could make the fertiliser itself much less polluting in both production and usage while still helping plant growth? Step forward CCm Technologies. They make farm-grade affordable fertiliser from waste products with around 80 per cent less carbon emitted in the total life cycle from production to use.

'If there was ever a time we needed this technology, it is now. We can maintain yield while slashing the climate impact,' says Peter Hammond who, along with Pawel Kisielewski, founded the company in 2011 after they met when dropping off their children at the school gates. I included their work in *39 Ways to Save the Planet*, but their science and engineering breakthroughs are so relevant to a book on land use that I wanted to catch up with them again.

'We have had huge success in terms of awards and backing from some very big food companies but we are a bit frustrated by the regulators' inability to ensure waste legislation keeps up with innovation,' Pawel tells me.

CCm combine three waste materials: CO_2 from chimneys; ammonia from food waste or sewage works; and fibrous (carbon-based) material, which could be more food waste, sewage solids or the 'cake' left over from anaerobic digestion plants. Using waste to make fertiliser may sound familiar because that is what we did for pretty much the entire sweep of human history before the Haber–Bosch process was perfected. We used our waste (night soil), animal waste (manure), food waste (swill) and nature's waste (leaf litter). CCm are diverting twenty-first-century waste streams to create a user-friendly and reliable product. Pawel is a fan of the Mahatma Gandhi quote: 'Waste is only a resource in the wrong place.'

They have had a long-term partnership with the food company PepsiCo at their Walkers crisp factory in Leicester. Here they have an anaerobic digester fed on waste potato peelings, which provides three quarters of the factory's electricity but also yields CO_2 and leftover digestate 'cake'. CCm Technologies combine them with further CO_2 from factory chimneys and ammonia from a nearby water treatment works to create fertiliser. This in turn is then used on the fields to help grow potatoes. Both the factory and the farming end up with much lower CO_2 emissions.

In late 2023 CCm signed a deal with the food giant Nestlé to supply 7000 tonnes of low-carbon fertiliser – enough to fertilise a quarter of the company's wheat harvest in the UK.

Another partner in this is the huge commodities company Cargill, who are supplying the organic material: cocoa shells left over from making chocolate. This follows a successful trial by Richard Ling, a farmer in Norfolk, who says: 'We've compared two parts of the field, one which used the cocoa shell fertiliser, and one which used the conventional fertiliser. There is no significant difference in the yield, so we can see that it works.'

Despite enthusiasm from some individual farmers, Pawel believes the key driving force will be big corporates, often demonised in the food system:

'When you are relying on Nestlé, Cargill and PepsiCo to drive the uptake it tells you something. It tells you that they are being incentivised to cut carbon by investors, regulation or by their own promises to cut scope 1, 2 and 3 emissions.* Whereas most farmers themselves are not incentivised to make these kind of changes. Another hurdle is outdated government regulation, which often makes it more difficult and costly to handle what is classed as waste rather than novel material. Even though we now know that, for the sake of less waste and less climate change, we *should* be using waste. The regulation penalises the virtuous.'

But their fertiliser has another benefit that they believe will cut through to farmers: it's better for the soil, and the dirty brown stuff is rapidly gaining respect.

* Scope 1 is what you emit yourself on-site by running heating, furnaces and vehicles, etc. Scope 2 is pollution made by someone else but directly commissioned by you, e.g. electricity to run your office or factory. Scope 3 are all the emissions up and down the value chain, including things like your suppliers, and this is usually where farmers come in.

'35 per cent of our product is carbon,' Peter tells me, 'which feeds and replenishes the soil. It releases the all-important nutrients more slowly in a more organically charged environment. It is, in effect, stuck within the lump of soil so we get 70 per cent less run-off, less leaching into the water on average. Also, the ammonium in our fertiliser breaks down more slowly so there is less loss to the atmosphere as nitrous oxide [a gas with nearly 300 times the global-warming potential of CO_2].'

Cutting those pollutants would be a huge deal, and, as I've mentioned before, it is this massive impact on our atmosphere and water that undermines conventional fertiliser's claim to be 'intensive' in the sense of limiting its impact to a confined space – untrue when its harm is so widespread.

Pawel and Peter are convinced that there are enough waste products in the world to supply most if not all of the world's fertiliser and turn feeding crops into a closed-loop system. The Carbon Trust – gold-standard climate-impact assessors – have just said that CCm's manufacturing process effectively removes emissions as it absorbs 0.9 tonnes of CO_2 for every tonne of fertiliser produced. They also reinforce CCm's claim that their product returns organic matter to the soil to help its regeneration.

CCm are not the only players in this space, as even the massive chemical fertiliser companies have pledges to reach net zero, but they are a great example of how science and engineering are relevant to land use: if we can do it smarter, we can do it with less.

10

BEHAVIOURAL CHANGE

What we do with the surface of the Earth is a product of natural resources and human skill on one side and human appetites on the other: supply and demand. So far, we have looked almost exclusively at the supply side. But changing what we want from our land would have equally profound effects. Hannah Ritchie at Our World in Data is one of a handful of experts I've talked to for this chapter on how our food choices – which are choices not necessities – affect land use and how altering our diet could change the surface of the Earth.

'Over millennia, we've just seen ever-increasing amounts of land being used to feed us,' she says. 'And we're now at the point where nearly half of the world's habitable land is used for farming. Most of that land is used for livestock. We've turned the world into a massive farm at great cost to biodiversity and great cost to natural ecosystems.'

Many observers point to the food system's unholy trinity that drives the land squeeze: overeating, meat and waste. Most of this book has been about satisfying current and future demands with the expectation of a similar diet: how we can work on the land to deliver more of the same suite of foods. This chapter is about demand and how that is shaped by our behaviour. To put it crudely: if we were all slim, vegan and never wasted a scrap of food, would farming pressure on land be released to such an extent that there'd be ample space for nature, energy, carbon storage and the rest? Perhaps the more difficult question is how any amount of this behavioural change can be achieved when food is so strongly bound up with pleasure, culture and identity.

Sarah Bridle is another star witness for this chapter. She is a professor at the University of York who specialises in how food affects the environment. Her PhD was in astrophysics but she decided there were more critical problems on Earth. Her book *Food and Climate Change Without the Hot Air* is a masterclass in robust yet accessible, even entertaining, data on the global-warming impact of our food choices. You want to know the carbon impact of your Victoria sponge, fish and chips, or flown-in French beans? It's all in Sarah's book. It is part of the 'Without the Hot Air' series started by the brilliant and much-missed David MacKay and opens with the phrase: 'Change your diet: the easiest way to help save the planet.' So I began by asking Sarah a very broad question: how important does she think our food choices are for land use?

'Hugely important,' she replies. 'Globally about 75 per cent of the agricultural land is used to produce food for animals. So, if we took an extreme case and went vegan, then we would not need that land. We would need a little more crop producing for humans because otherwise we'd go hungry. But it takes on average about sixteen times as much land to produce a calorie of animal-based product as plant-based product.'

This figure combines both grazing land and cropland delivering animal feed.

'Because there's this factor of sixteen,' she continues, 'in the extreme case, which I am not advocating, if the whole world went vegan then you would need an extra 5 per cent of the agricultural land for producing crops to make up the calories. But then that would free up the majority of agricultural land for other uses including nature-based solutions. There is huge potential in there.'

Of course, calories aren't the only important ingredient in food; protein matters too, and livestock advocates are keen to point out that meat and milk are an excellent protein source. But here again vegetables, like grains or peas, need a fraction of the land area to produce a gram of protein compared to meat and dairy. The only animal products that begin to rival plants in the space race are chicken and farmed fish.

Hannah Ritchie sees an upside in the calorific conversion inefficiency and general wastefulness of our food system. She says it could give us both resilience to short-term shocks and room to make long-term changes:

'Many people predict there's going to be a food collapse, that we are just at the whims of nature when it comes to food. The key point I want to get across is that a lot of this is in our control. People have no idea how much food the world produces. Because there are around 800 million people in the world that don't get enough food to eat and there's many people that eat a little bit too much, they think that probably evens out and we produce just about enough per person. Whereas, in fact, we produce more than 5000 calories per person per day [around double the average human daily need]. There's a massive loss across that food chain. One is just standard food losses – waste. Another is the amount we feed to livestock, which is very inefficient. Then there is the food that we put into cars, in the form of biofuel, which is not a good use of that. What I want to get across is that we have massive agency in the system. It's not that we are just on the edge of not being able to be grow enough. We produce more than enough; it's about what we do with it. And what we do with it is completely within our control.'

So, I ask her, do we have enough land for all the things we need it to do?

'Yes, absolutely. But we won't get there without massive changes. We need to eat significantly less meat. We would need to stop using biofuels for cars and we would need to see continued increases in productivity of crops. But I think that, technologically, it's more than possible, so it becomes a choice of whether we actually do it.'

Helen Browning, who has a farm with pigs and dairy cows alongside running the Soil Association, is also in no doubt that meat consumption is too high.

'If the rest of the world all eats as much meat as we do in the UK or the US,' she tells me, 'we blow the planet. We can't do it. And I'm particularly nervous about where those livestock compete directly with humans for food and feed, the way we feed the animals that feed us and the amount of grain and protein that is diverted into the livestock sector.'

Most studies – including those commissioned by government – agree that, when it comes to meat, 'less would be better'. Whether 'none is best' remains a moot point. While climate, land-use and animal-welfare purists might back total veganism, it is perceived as impractical, political unachievable and, if not potentially unhealthy, at least nutritionally complex as it is difficult to get all the required nutrients from a totally plant-based diet. The UK National Food Strategy (sometimes referred to as the Dimbleby Report after its main author, Henry Dimbleby) was commissioned by the government in the early 2020s. It stated very clearly that, in order to meet climate, nature and health commitments, meat consumption should drop by 30 per cent by 2032. It suggests this would allow space for the creation of 410,000 hectares of woodland, restoring 325,000 hectares of peatland and managing a further 200,000 hectares of farmland mainly for nature, like species-rich grassland with some conservation grazing. That is a total of 9350 km^2 made available for climate and wildlife priorities – an area just shy of half the size of Wales – from eating almost one third less meat.

The Climate Change Committee (CCC) is the independent body set up by the UK government to advise on how to meet its climate commitments, the so-called carbon budgets. It too is quite clear: 'The CCC recommends a 20% reduction in meat and dairy by 2030 and 35% reduction for meat by 2050.'

Before we come on to if and how such cuts could be achieved, let's take a look at our waste and our waist. It is commonly said that one third of what is grown is never consumed. It is wasted on the farm, in transit, in the shops, in restaurants and in our homes. The pattern of where food is lost varies globally. In the economically poorer world, the lack of efficient logistics and a reliable cold chain (refrigerated transport and storage) means much food spoils before it reaches the kitchen. In the richer world, the bulk is binned in our homes and eateries. The consequent 'waste' of land is huge. Without any food waste, we could let one third of our farmland – an area the size of Russia, the world's biggest country – return to nature.

These losses aren't simply a waste of the space that grew that discarded plateful, they are also environmentally punishing. Growing food produces greenhouse gases, and its rotting releases more. Around 8 per cent of human-induced climate change comes from our food waste – about the same as emissions from India and Germany combined. A zero food-waste world would leave more land to plant carbon-sucking trees and less rotting food emitting greenhouse gases. Combine the two and you have a big bite out of climate change and more space for nature.

Waistlines are growing too. Data from the United Nations Food and Agriculture Organisation shows worldwide calorie intake per head has grown nearly 10 per cent in the last twenty years. The growth is steepest in Asia, but increases are also seen in Africa from 2400 to 2600 calories per day. In North America and Europe the figure rises from a more belt-busting 3450 to 3550 calories. We should not forget that close to 1 billion people are still under-nourished and that enough calories does not mean a healthy diet. In much of the richer world we are now experiencing the odd situation of more obesity in poorer communities (a vast subject beyond the remit of this book but linked to availability of cheap processed calories and inaccessibility of healthy food and nutrition education). Overall, there is no doubt that the overfed now vastly outnumber the underfed and eating too much means growing too much and using too much land.

And there is the problem: we are greedy, squandering, carnivores. Our overeating and meat-eating appear to be escalating as the world gets richer. With the world population also rising, these pressures on land use seem to be increasing, so what are the solutions?

Dr Tara Garnett is based at the Environmental Change Institute in the School of Geography and the Environment at the University of Oxford and is co-investigator on the Livestock, Environment and People project. She is also director of TABLE, the global platform that analyses values and viewpoints on food-system controversies and she outlines the spectrum of four responses to the diet and land-use challenge identified by TABLE.

'One says "we are where we are, you can't do anything about demand", therefore what we need to do is produce more with less,' she says, 'and in order to do that we use land as efficiently as possible. So, for instance, when it comes to livestock we move to more monogastric consumption [pigs and chickens, not cattle or sheep] – the world likes poultry anyway as it doesn't offend many religious sensibilities. This is production-side sustainable intensification. Next, we have a super-charged version of the above, which bypasses the whole animal-production process and invests very heavily in alternative proteins [fake meats]. This avoids all the problems with animal sentience. There is a strong environmental motivation, but I think for many it comes from an animal ethics, animal justice perspective. This is still a kind of technological fix. The third option says "no, you've got the whole thing wrong. You have to think about the traditional role of livestock in the food system." This is what I call the "livestock on leftovers" approach. They are resource recyclers and can exist on land unsuited to other purposes. So rather than thinking about the quantity of meat, you are thinking about quality. In the fourth scenario, you cut out animals altogether, base your diet around legumes, pulses, grains and vegetables, and give the land unsuited to crops over to rewilding.'

Tara wants me to know she is a vegetarian and that this could affect her views, but she says it's very hard to see the consume-as-usual scenario with techno fixes on the farming side working in a world of 10 billion people. She doesn't want to close the door on novel resource-efficient food solutions but believes they are a long way off. Therefore, smaller amounts of meat in our diet are a big part of her solution but, she tells me, 'There has been an inverse relationship between measures that the government are prepared to take and the effectiveness of those measures.' This is academic speak for saying policies tried so far in the UK don't work. She believes this is mostly because these policies usually put the responsibility on consumers (governments would call this offering choice), but these shoppers exist in a food environment that is geared away from encouraging sustainable or healthy diets.

'I would make a strong case for the fact that governments and the food industry must step up, signal a vision of where they want to go and then create that enabling environment,' Tara argues. 'Once you get people talking and understanding the issues, they are much more open to the idea of government playing a role in our diets, particularly for their children's sake, than the myth of the indignant public would have you believe. I think industry and retailers should be a bit bolder. They're happy to sell you fake meat but they are less happy about creating less shelf space for actual meat. They like more product diversification. But whether that leads to less [meat sold] is another matter. I could see a role for a meat tax, especially if combined with hypothecation [separating out money raised through taxation for a specific use] that subsidised healthier and more sustainable foods.'

Hannah Ritchie believes that some degree of behavioural change is required across many aspects of our lives to deliver an environmentally sustainable future but that it's particularly difficult to swallow with food: 'I'm much more optimistic about the energy transition than I am about the food transition, because I think the energy transition is much less hinged on personal behaviour change.' When you turn on the kettle, you don't know or care if the electricity came from a wind turbine or a gas turbine, she says, but: 'Some people have this impression that if you just tell people to switch to beans and lentils then they'll just do it. And that's just not going to happen.' She thinks dietary change will only come when it is as invisible as the revolution that happened behind your light switch – a technological fix: 'The only credible path I see is producing alternatives that exactly mimic the experience of eating meat. There I could see a credible path where people would just make the switch naturally.'

Helen Browning definitely favours keeping real animals in welfare-friendly systems but fewer of them. She describes this as 'less but better' at the consumer end and 'livestock on leftovers' for the farmer. Both should result in a lower-meat diet. Meat should be produced largely outside, on land that is 'leftover' (i.e. unsuitable) from arable production. Supplementary animal feed, where required, should come from unavoidable waste from the human food chain. Intensive livestock farming is out and so is cheap meat. But this approach has a couple of challenges for both climate and equity.

It is widely acknowledged that intensively reared animals have less climate impact per kilo of flesh than those skipping around the fields. This is because they grow fast, die young and don't waste calories wandering about. Factory-farmed chickens in the UK now reach their slaughter weight in forty days or so and each get a minimum area of ground about the size of an A4 sheet of paper. They are the unlucky icons for this 'efficient' form of farming, but similar inconvenient truths apply to the rearing of pigs and even cattle: low carbon often overlaps with low welfare. But Helen says we cannot farm under a climate-metric tyranny: 'The carbon equation of a free-range pig is less good than one that is incarcerated. But I just don't think we can go there. You can't justify inhumane systems on the basis of a lower carbon footprint. We just have to say "let's not do it".'

She can see a role for artificial meat at the cheaper end of the market – 'replace the chicken nugget with precision fermented protein and add flavouring' – but essentially believes that we shouldn't be embarrassed about good meat being expensive. We expect to pay for quality in every other part of our lives: 'In many other markets we don't have that same disparaging attitude to paying more for anything that's better. If you aspire to have a nicer car or a bigger TV and it costs more, no one says that's a simply dreadful thing. In other areas, we accept that better costs more and we don't seem to accept that in food.'

Helen is very clear that a good diet with occasional meat can and should remain affordable with sufficient availability of healthy foods and knowledge of how to prepare them. But it is hard to avoid the conclusion that, in this scenario, good meat should be a luxury – and wealthier people can afford more luxuries more often. Although essentially true, I can't see that being a popular political slogan.

It seems to me that eating less meat is a smart way to reduce pressure on our land. The figures underpinning it are solid and the health arguments (with the exception of some communities in poorer countries, who don't get enough protein) have persuaded one of the most unlikely converts. I interviewed film star, body builder, politician and green campaigner Arnold Schwarzenegger and asked about the place of behavioural change in cutting greenhouse gas emissions. 'I have reduced meat by about 70 per cent,' he told me. I suggested this must have represented giving up something, a personal sacrifice, but he replied: 'No, I gained something. I gained my health. My doctor says I've added at least another two years to my life. Now I just have to give up cigar smoking!'

Away from movie stars with their own personal physician, I suspect the way to reduce meat is a combination of all of the above, which can be summed up as making meat less attractive and plants more so, whether through price, tastiness or social pressure. In my own diet, I have become more plant-based as I've tried more good veggie recipes, have grown more greens and become more aware of meat's harms. But no one has forced me to do it. I love food, it is a source of great joy and companionship and I hate the rising tide of guilt at the national dining table. My food choices are mine with inevitable influence from my social and economic surroundings.

When it comes to behavioural-change success stories, people often cite anti-smoking, where law, tax, advertising and public scorn combined to slash the number of smokers. Some ask why we can't do the same with meat consumption, but I think it's a bogus comparison. Smoking performs no useful function – apart from making film noir screen gods look devastatingly cool – and, compared to diet, its historic and cultural roots are shallow. Meat is, in essence, a very high-quality foodstuff. From the Sunday roast to the chicken-shop takeaway or the 'full English', it is linked (though not bound) to our identity. Meat eating can't and shouldn't be demonised in the same way as something that mainly causes heart disease, halitosis and cancer.

Although I'm not in favour of a full onslaught on carnivores from the arsenal of government policy weapons, I would favour a clear statement that eating less meat is a desirable future. But how do we get there?

If meat eating divides opinion, food waste unites it. No one thinks food waste is a good thing; everyone thinks it should be reduced. In France such broad support allowed them to pass a food-waste law that obliged major retailers to donate almost all unwanted food to charities. However, it's within our own four walls that most of the food waste meets the bin. UK households bin food worth over £14 billion per year, and further millions are wasted in shops, restaurants and takeaways. Why? I have a simple answer that is so unpopular it is almost off limits to politicians and campaigners: food is too cheap. This may sound like heresy at a time when food banks are in great demand and food inflation has recently spiked. Yet most consumers can afford to waste because we spend a much smaller proportion of our income on food than we used to. According to the UK government's National Diet and Nutrition Survey, 24 per cent of a typical pay packet went on food in 1974 compared with 11 per cent today. While it is true that people on lower incomes spend a greater proportion on food, that figure too has declined. The economic pain of binning a curling, drying fridge lurker is less than the effort and inconvenience of using it up, especially when those freshly cooked alternative options can now be delivered straight to your door.

I asked each of my experts for this chapter pretty much the same question: is it possible to cut food waste substantially while we insist food is cheap?

Food academic Tara Garnett replied: 'Personally I don't think it is. Food being very cheap, having a fridge and 24/7 opening hours all work together for you to put that food in the back of the fridge and forget it. So you kind of waste food not in spite of your fridge but almost because it's there.'

Statistician Hannah Ritchie's immediate response to this was a very long pause followed by a pretty direct answer: 'No, probably not. I think changing consumer waste is very difficult. I think food needs to remain cheap and that just comes at the cost of food waste.' She believes food waste is inevitable collateral damage from a *desirable* cheap food culture.

Organic food champion Helen Browning shares this analysis but not the conclusion: 'People waste less when food becomes more expensive, and the system is perversely generating loads more waste because it's just too easy and too cheap to do so. The ultimate incentive to waste less is that it is too valuable to waste.'

Helen also dares to question the unquestioning damnation of all food waste:

'I sometimes push back on the idea that waste is the biggest bogeyman in town. We have to think about a degree of redundancy in the system that allows us to deal with shocks and instability. How do we actually make sure there is enough food in the system when we're dealing with a really uncertain climate? In fresh produce, for instance, while it feels scandalous that you end up ploughing in your lettuces… sometimes you might need a little excess in the system, given that we really do not know how each season is going to turn out. Trying to get that balance of supply and demand absolutely spot on, so that there is no waste, means that you're likely to end up with shortages at times.'

In effect, complete food security requires some element of food waste.

The final 'bad habit' we should kick in order to spare more land from farming is eating too much. The Our World in Data website displays our growing appetites starkly: between 1975 and today, world obesity rates have gone from less than 5 per cent to around 15 per cent. Over the same period, you can move a slider along the years and watch different countries turn from white to red as their populations put on weight. Graphs show clearly that growth in wealth and growth in weight are very strongly linked: we get richer, we get fatter. Only south and east Asia (mainly China and India) seem somewhat immune to this trend.

Governments have dared to be bolder in fighting fat than meat, not due to the land-use impact but because the consequent ill health harms people and health-service budgets. In the UK we have mandatory calorie labelling, traffic-light colours warning of high quantities of sugar or fat, and most recently a sugar tax. In 2017, this imposed an 18 pence per litre levy on soft drinks with 5–8 grams of sugar per 100 ml and 24 pence per litre on those containing more than 8 grams per 100 ml. (There are some exemptions for fruit juices and milk-based drinks, which are thought to have some nutritional upsides.)

The most obvious result of the tax was a reformulation of drinks to bring their sugar content under the threshold, and total sugar sold in soft drinks decreased by around one third in the first five years of the tax – that's 42,000 tonnes less sugar. The effect on waistlines is more disputed, with some studies suggesting less childhood obesity in girls, but the overall levels of obesity in both children and adults continue to increase. This doesn't mean the tax was a failure, as we don't know how much greater the increase would have been without the tax – there is no 'control experiment UK' where we can see the result of no intervention. Our body weight is, of course, a product of both consumption and exertion, so a greater body mass index doesn't always mean more calories.

So where does all this leave the claim the changing your diet is the easiest way to save the planet? Is behavioural change the easy, greased lever just waiting to be pulled? In one sense, yes. I have the final say over what I put in my mouth, and that decision – if taken by millions – would have a huge impact on the land area required for food. But we live in a massively influential and culturally entrenched food system that thrives on selling us more. We all know changing diet is challenging personally; it is for governments too. Our land-use challenges *could* be solved by altering food choices, but relying on that is wishful thinking and waiting on a train that ain't gonna come. Unless it is forced upon us.

Professor Tim Benton is an ecologist turned food, politics and environment expert based at the international affairs think tank Chatham House. From 2011 to 2016 he led the UK's Global Food Security Programme. He says: 'What is politically acceptable today is not the same as what was politically tractable five years ago. It won't be the same as what is politically tractable in five years' time.'

He observes that we live in an increasingly unstable world, where nationalism is on the rise and transnational cooperative institutions – like the European Union, the

United Nations and the World Trade Organization – are under siege. And you can layer climate change on top:

'If you think to where we were ten years ago, Brexit wouldn't have been imaginable. Trump would not have been imaginable. A shooting war in Europe wouldn't have been imaginable. We have increasing fragility, increasing contestation and increasing volatility of events at a global level. Wherever you look, there's more political division and potential for all sorts of nasty things to go on. And the speed with which extreme weather is evolving would only be imaginable to a few nerds, which is partly why we're in that situation.'

He suggests that the current cheap and plentiful food system relies on smooth international trade, which is looking extremely vulnerable. A shock to that system would demand change: 'We might well have to be producing more food for local consumption. If we're producing more things for local consumption, can we afford the current balance [high consumption of fertiliser and meat]? And the answer is no. We would have to have a more diverse system. We'd have to think about having higher-tech but more mixed diverse farming systems, integrating livestock into arable because we wouldn't necessarily be able to get hold of the nitrogen fertiliser produced by countries that we don't want to trade with, in the same way that we don't get gas from Russia anymore. So you can imagine a world where events in combination with the politics spur new spaces for decisions that at the moment feel politically unpalatable.'

History suggests that wars drive rapid food-system changes (such as food rationing and the 'Dig for Victory' campaign during the Second World War), and it seems plausible to me that security threats could drive more efficient land use. However, this isn't a choice but rather a side effect brought about by terrible human suffering, and certainly not something that anyone, least of all Professor Benton, would advocate. But he thinks another looming crisis – the one in our health systems – could embolden governments to influence what we eat.

'They're at breaking point,' he says of our health systems, 'because of obesity and ill health, and that's not going away. It's just politically unpalatable at the moment, particularly for the right wing, to say "we're going to encourage you to eat healthily" because the *Daily Mail* and the *Telegraph* will see that as nanny state-ism. At some stage we're going to have to deal with that as we won't have the money to run a health system where everybody is ill. You can see many governments, even in the global south, starting to grapple with these issues: that if you just leave the market alone, the job of the market is to sell things for profit. And if they can get away with selling you unhealthy foods at volume and make you want to eat more and more and more of it, they make profits and then there's a public health consequence and that's covered by the public purse. You can only do that for so long before something has to change.'

Eating patterns have changed in the past and may change again in the future. Given the persuasive health and environmental arguments for less meat, less waste and fewer calories, I think a confident and wise government would spell out the facts and nudge us in the right direction. I am not a vegetarian or vegan and I don't favour a meat tax because I don't think *some* meat is a bad thing. If a 'frequent meat-eating tax' was remotely plausible I could support it, but it isn't.

The conclusion is that behavioural change is an important tool in the box for delivering smarter land use, but it is not to be relied upon for completing the job.

CONCLUSION

So far, this book has taken you on deep dives into the realms of thoughtful and innovative farmers, scientists and even warehousing managers, all with a passion for efficient land use. To end it, I'm going to take a bit more of a helicopter view to reveal the prevailing strategies that my case studies have in common. It will also place my mainly British examples in a global context. As a spine for this, I'm going to use the World Resources Institute's four pillars for addressing the land squeeze and the associated crises in nature and climate: Produce, Protect, Reduce and Restore.

According to the WRI website, 'Rather than create an activist organization, its visionary founder, Gus Speth, established a science- and evidence-based institution that would carry out rigorous policy research of global environmental and development issues.' I like the absence of overt ideology, the rigorous science base, and their mission: 'Moving human society in ways that protect Earth's environment and its capacity to provide for the needs and aspirations of current and future generations.' So let's take their four pillars one by one.

Produce

The population of the world continues to grow and may reach 10 billion by 2050. That is an extra 2 billion people. More mouths means more food, and it must come from the same farmed area. The WRI's solution can be summed up as benign intensification.

For meat production they recommend better pasture, rotational grazing (where the grass is left to grow longer and the stock is moved more frequently), silvo-pasture (trees among beasts), increased amounts of nutritious feed in stall-fed systems, improved breeding and better vet care. The arable menu includes new crop varieties that are higher yielding and more resistant to climate change, as well as better water and soil management, slow-release fertiliser and chemical usage with less collateral damage to wildlife, expanded agroforestry and more farmer training.

This list is focused on spreading the best practice to deliver high productivity with limited environmental damage, but it isn't especially radical (apart from a love of trees), and conspicuous by its absence is the term 'regenerative farming'. The WRI's managing director of Strategy, Learning and Results, Janet Ranganathan, says: 'It's hard to have a precise view on regenerative agriculture because it is really a collection of different practices. It's a term that has been used for organic agriculture, right through the spectrum to more industrial agriculture... It's an unhelpful term until it is defined.'

I can understand how what I see as open and welcoming would be seen by an academic as frustratingly vague. But

Janet has another concern, with some justification, that regenerative farming often ignores the impact farm-scale decisions will have on the global-scale food system.

'For example,' she says, 'if a practice reduced yields by 15 per cent then the system consequences of that may be that food prices would go up or it would drive expansion of the agricultural frontier. That could lead to converting natural ecosystems that store large amounts of carbon, and that would more than offset the original benefits at the farm level.'

This is a conundrum that runs through this book and why I am championing those smart farmers who care about both food production and environmental health. I have come across convincing examples of this under the regenerative banner and I firmly believe that achieving this should be the focus of smart land use. Janet's scepticism is borne of experience in seeing well-meaning farmers believing they are doing the right thing by prioritising nature while ignoring the collateral damage of less food production. But surely we should be equally harsh on chemically intensive farmers who are ignoring their side effects on nature, climate and local pollution.

Protect

We need to protect all the remaining wild land – not just forests, but wetlands and grasslands too – with strong laws backed by sharp teeth. This is an ambition supported by the Intergovernmental Panel on Climate Change and a wide consensus of politicians, conservationists and scientists. The World Database of Protected Areas records 287,359 sites across 244 countries and territories, which together cover 17 per cent of the world's land and inland waters. But designation is a lot easier than protection. Much deforestation is already illegal but happens anyway because the law is weak or poorly enforced. According to the United Nations, at least 70 per cent of the tropical forest cleared for agriculture between 2013 and 2019 was done in violation of national laws or regulations.

We need to build an impenetrable force field to guard our natural world piece by piece: elect politicians both locally and nationally who will prohibit natural ecosystem conversion and put hefty penalties on lawbreakers. Grant greater rights to indigenous people to manage their own land. Plan roads and railways to avoid cutting through and therefore opening up virgin territory. Work with businesses to create supply chains genuinely free from the conversion of pristine land. Increase knowledge and transparency of businesses operating on the wild frontier and the actual data on land use. Finance the maintenance of untouched land by paying for its ecosystem services – like flood prevention and carbon storage – or promote livelihoods, like ecotourism, that make money from the wilderness being kept largely intact. This last method has been pursued most famously by Costa Rica, which now earns billions of dollars through ecotourism. Other countries like Rwanda and Benin are taking note.

The overall picture for tropical forest remains bleak as the tree-loss data for 2022, compiled by the WRI with Global Forest Watch, shows deforestation rates increasing

– despite international commitments – especially in Ghana, the Democratic Republic of Congo, Bolivia and Brazil. It is too early to tell whether the election of Lula da Silva in Brazil in late 2022 – a president committed to ending deforestation – will reverse this trend. However, there are some countries who seem to be getting protection right: most notably Indonesia, which now has lower levels of forest loss than in recent history. Malaysia, China and Costa Rica have also reduced tree-felling levels by more than half in recent years. Indonesia has a target of being a 'Net Sink' – negative carbon emissions – by 2030 and has introduced more fire prevention, ended exploitation licences of primary forest or peatlands, established better law enforcement and replanted mangroves. Indonesia's corporate pledges to reduce the deforestation associated with palm-oil production may also be having an effect.

What about specifically protected parks? Janet believes they only do some good if they are part of a wider strategy: 'If you just put in that park and don't have a strategy for meeting the demands that are driving [farmland] expansion, all you do is displace the problem. So it is better to look at the landscape or system level; otherwise you might just be dealing with a lot of leakage [of farming into unprotected wild lands].'

A growing and potentially powerful land protector is carbon finance. One of the options that polluting companies or even countries can pay for, with a view to offsetting their emissions, is protection of large areas of vulnerable forest. This is known as 'avoided deforestation'. It only has merit if two criteria are met: the forest is genuinely at risk and the protection will be assured for many decades. I've seen examples from Brazil where satellite photos of vast chunks of land show strips of cleared land encroaching on large patches of pristine forest. Payments to local communities and enforcement go towards ensuring the forest's ongoing protection. There are many ideological and practical questions about this approach – not least, whether it encourages countries to deliberately threaten wilderness just so they can get money to save it – but I've met plenty of very committed people working in this field who reckon they are doing as much to 'save the rainforest' as a protester on the streets. At the very least, carbon finance is emerging as a multimillion-dollar pot of money earmarked for protection.

The WRI has another, less familiar candidate for protection too: highly productive farmland. The threat here is frequently from housing as many of our cities grew from settlements in the most fertile farmland, frequently wide river basins with rich flood plains. As urban areas continue to grow, they often pave over the best soil. I toyed with dedicating a whole chapter to housing in this book, not least as urban sprawl and new housing estates are so visible and political, but urban architecture is well beyond my natural habitat. My amateur prescription would be to make high-rise, high-density living appealing and to build nature into cities wherever possible. Also, in terms of total area around the world, housing is not seen as a major driver of land-use change. Yet that doesn't mean it should be ignored, especially as land delivering bumper harvests is often at risk. A study published in the *Journal for Global Environmental Change* estimates that 65 million tonnes of crop production will be displaced by urbanisation in the period 2020–40. This in turn could drive the development of 350,000 km^2 of new arable land by taking it from either pasture or wilderness. The problem here is how we value land: for food it is worth very little, whereas for housing it is worth fortunes. Most urban edge farmers dream of getting planning permission on a field or two, so it is highly unlikely the market will deliver this protection. Instead, it will be up to law and planning guidance. As the WRI says, 'To address the land squeeze, highly productive lands need to stay highly productive.' I am sure those fighting to keep solar panels off their local farmland might feel the same way.

Reduce

We've looked at the supply, now what about demand? Changing our habits to consume less food that demands a lot of space would free up so much land for wildlife, carbon storage, recreation or all of the above. This takes us into the realm of behavioural change, as we saw in chapter 10. Unsurprisingly, in common with most experts in the field, the WRI concludes that we can't be smart about land unless we are smart about food.

Reminding us that pastures used to produce ruminant meat (cows, sheep and goats) or their feed comprise two thirds of all agricultural land, the WRI says we should: 'Shift diets of high meat consumers toward plant-based foods. From a land and climate perspective, a critical component of this measure is reducing growth in consumption of meat, especially beef. This entails increasing the supply of meat alternatives including plant-centred meals, plant-based foods that mimic meat and lab-grown meat.'

For a summary of sage food advice, it's hard to avoid the straightforward clarity offered by American journalist and author Michael Pollan: 'Eat food. Not too much. Mostly plants.' But reduction is not just about the stuff we eat, it's about the stuff we burn: biofuels. We've already had a hefty swipe at them being a really inefficient use of land in chapter 3, including a contribution from Ed Davey at the WRI, but I'm going to throw in a couple more WRI statistics here to clarify quite how land-hungry biofuels can be: 'To produce just 2 per cent of the world's energy through liquid biofuels would require agricultural production to increase by another 30 per cent by 2050.' That's a third more land under the plough for that tiny gain. There is more: 'Even on the best lands for bioenergy, a hectare of modern solar photovoltaics can generate approximately 40 to 100 times the amount of usable energy as biomass.'

Janet Ranganathan says there could be a small role for biofuels from waste products that could not be used for anything else, so I ask if she has ever come across a benign form of biofuel. 'No, not at scale,' she replies. The promotion of biofuels through government regulation or subsidy needs stopping as soon as possible, although it should be remembered that in many countries firewood and charcoal are still the primary energy source for cooking and sometimes heating. This is bad for health as indoor air pollution through smoky hearths worsens lung conditions, heart disease and many more illnesses. Where possible, these heat sources should be replaced by small-scale solar or cleaner fuels – even if that means those derived from fossil sources. Protecting the local woodland and wellbeing of the household is better for humanity and probably the climate too.

Restore

How can we restore land that is degraded or abandoned so that it delivers some of those natural and public goods we want? The WRI favours a range of approaches, including assisted natural regeneration in which the land is left alone for the small pockets of native plant populations to expand with interventions only to curb invasive species and reduce overgrazing. A good example of this would be some Scottish Highland estates where little is done apart from having the sheep removed and the deer fenced off or culled, and then the patches of Caledonian pine or indigenous scrub are allowed to expand without the constant grazing pressure.

Another approach is active restoration, which involves bringing back natural vegetation with extensive planting, seed collection and nursery management. Such projects often require long-term maintenance to avoid the return to non-natural degradation. Replanting mangroves on tropical coasts once damaged by aquaculture or industry would fit this approach. The WRI is pretty explicit that some areas are not right for restoration, such as good or even average farmland, 'because restoring ecosystems in one place poses a high risk that other ecosystems will be converted to make up for the foregone production'. An exception to this would be peat, where the greenhouse-gas emissions from farming are so great that the food could well be grown with less harm elsewhere.

On restoration, it is worth taking a detour into the United Nations' Convention to Combat Desertification (UNCCD). They work in a similar way to the global climate or biodiversity summits in bringing together as wide a group of national leaders as possible along with scientists, private companies and NGOs to agree on ways to tackle transnational problems. In 2021 the UNCCD and its members made a commitment to 'restore 1 billion hectares of farms, forests, and pastures – an area greater than the size of the United States or China' by 2030. They calculate this will cost around $1.5 trillion, which is much less than current subsidies paid to the farming or fossil-fuel industries.

This is all about restoring degraded land, which – in the strictest sense – is an area of the Earth's surface that has lost some degree of its natural productivity due to human intervention. But this broad sweep would include just about all land touched by farming, building or land management. In essence we are talking about land that has lost the capacity to do much of anything: produce food, store carbon, house wildlife, house people, move people, prevent flooding – basically all the things this book suggests land is for. Common symptoms of land degradation are soil loss, desertification, high carbon emissions, little life, pollution and declining food yield. It has usually been caused by land-use practices, such as tree felling, overgrazing, drainage and excessive use of fertilisers or biocides. It's easy to be damning of these behaviours, and we certainly need to think again about them, but we could also look at that list of evils and say 'many of those are keeping us fed today'.

The four pillars of Produce, Protect, Reduce and Restore are not a menu where you can choose your favourite. They are a recipe in which all the ingredients are required to produce the desired dish. There is no point in increasing productivity without protection as that just encourages more profitable production to grab more land. Innovation is essential too – not just in novel crops, climate-friendly cattle or cheap and tasty veggie 'meat', but also in finance to develop models that better value nature or in social science to sweeten behavioural change. And these things go well together: I am more likely to accept policies to drive down meat on the plate if the alternatives are affordable and delicious. I am not saying the WRI's ideas have the monopoly of virtue (for instance, I am much more enthusiastic than them about the potential of regenerative agriculture or commercial forestry) but I feel they have had a clear-eyed look at the huge overlapping challenges and, based on analysis not ideology, come up with a convincing blueprint of what's needed – like space-efficient farming, biotechnology, less meat, and rewilding uplands – even if some might find them hard to swallow.

Throughout this book, I have admired those who have been able to deliver multiple uses from the same piece of land. But working towards multiple goals is complicated, whereas a single priority is simple. It requires you to understand trade-offs and mutual benefits and may require very diverse skills – from soil science to solar-energy engineering. Farmers understand multi-tasking as they are often a one-man or one-woman band, but a land-smart future demands this at another level.

For too long, land-based skills have been somewhat looked down upon by the academic establishment all the way from school through to higher research. As an under-graduate and part of a middle class with university-educated parents, I remember assuming a Land Economy degree was second-rate. This must change because smart use of land is essential to our future, and vital to that is education. Janet Swadling had leading roles in Scotland's Rural College, has received an OBE for services to education and is now director of the newly formed Institute for Agriculture and Horticulture that has been created to upskill those working with the land and to recruit more clever entrants.

'The Institute had a stand at a careers fair at the University of Cambridge,' she says, 'and it was all Master's students who'd done subjects like Geography or Environmental Science and, almost without exception, *none* of them had thought about a career in agriculture. They just hadn't seen it being for them but were now saying, "Oh yeah, this is where I could really use my skills to make a practical differ-ence with climate change." It's a lot more "sexy" to go into cancer research or computing. But land-based skills are "save the world" skills. It is vitally important that we have the brightest people producing the best quality of food that we eat and [enhancing] the environment as it has a direct impact on people's health and the health of the planet. Goodness knows we need to do something about that.'

Our one planet can deliver all the needs of humanity and the natural world but only with the intense application of intelligence and very little dogma. This book showcases people who are doing just that. Spread the word.

ACKNOWLEDGEMENTS

Caroline Drummond MBE, to whom this book is dedicated, must come first. She ran LEAF (Linking Environment and Farming) for over thirty years from its foundation in 1991. The organisation and its hundreds of farming members exist to show, through demonstration and science, that sustainable commercial agriculture is possible and desirable. Their Leaf Marque, requiring a robust environmental audit, is now on food from 310,000 hectares of farmland worldwide and their Open Farm Sunday event has welcomed 2.7 million visitors through farm gates since 2006.

Caroline's style was unfailingly clear, frequently brave and always pragmatic: based on what works, not led by ideology. She was the first person I spoke to for this book, as over the years, we had become friends and advisors to each other. Under her name in notebook #1, the first word is 'head-bending' then 'good luck', followed swiftly by suggestions of places to visit and other people to talk to.

Caroline died of cancer aged 58 in May 2022. Her enormous legacy lives on. Now with the modest addition of this book.

Alongside those featured in this book, I also owe huge
thanks to those I spoke to in order to frame my thoughts.
As the book is dominated by eye-witness accounts much
of this influential counsel goes uncredited: conservation
scientist Professor Andrew Balmford; biotechnolo-
gist Dr Tina Barsby; Shaun Spiers from the Green
Alliance; renewable energy pioneer Juliet Davenport;
Craig Livingstone from the Lockerley Estate; Jonathan
Wadsworth from the World Bank; Daniel Pearsall from
'Science for Sustainable Agriculture'; environmental
economist Professor Ian Bateman; Chris Buss from the
International Union for the Conservation of Nature
(IUCN); sustainability economist Toby Gardner; regen
farmer Paul Cherry; farmer and journalist Tom Allen-
Stevens; Oliver Blair from Dyson Farms; environmental
scientist Professor Jules Pretty; Louise Baker United
Nations Convention to Combat Desertification.

From the publishers, Atlantic Books, editorial director
James Nightingale has shaped the narrative and been
simultaneously patient but no pushover in the face of my
constantly distracting broadcasting commitments. Charlotte
Atyeo polished my prose and picked up some shocking
geographical errors. Thanks also to my book agent Patrick
Walsh and Derek Wyatt, the man who first brought me to
the attention of Atlantic.

My best friend, Martin Evans, kept my author's hubris
in check, and my wonderful wife Tammany is my ever-
present sounding board and gives unflinching support to an
overworked old hack.

INDEX

Abbey St Bathans, Berwickshire, 174, 177–88
Aberdeen, Scotland, 93
Aberystwyth Plant Breeding Station, 131
accidental agriculture, 33
afforestation, 6, 39, 188, 192, 194–206, 290
Africa, 15, 82–3, 232–3, 268, 285–6
agronomists, 141, 144–9
Aiden, Saint, 103
air freight, 156
Aire River, 216
alder trees, 196, 201, 203
algae, 105–6
allotments, 17–20, 22–3, 35–6
Alqueva, Portugal, 82
alternative proteins, 270, 271, 272, 273, 292
Alzheimer's disease, 90
Amazon region, 209, 240
American Land Institute, 249
ammonia, 103, 104, 105, 260, 262
anaerobic digestion, 98–9, 160, 260
animal feed, 9, 15, 138, 264, 265
Anne, Princess Royal, 255
Antarctica, 209
anthelmintics, 253–4
aphids, 224
apples; apple trees, 21, 175, 201, 248
arable farming, 102–22
    climate crisis and, 112–13
    fertilisers, see fertilisers
    intensive, 114, 224–5, 243, 262, 283
    soil, see soil
    wildlife and, 111–12, 114–15

Arable Innovator of the Year, 122
Argentina, 40
Armstrong, Alona, 54–5
artificial meats, 270, 271, 272, 273, 292
ash trees, 134, 140, 204, 217, 218
Asia, 15, 84, 95, 268, 277, 286
asparagus, 156
aspen trees, 196
aspirin, 90
Attspodina, Karolina, 76
Australia, 175, 209, 241–2

bacteria, 28, 110, 116, 120, 146–7, 166, 180
Bala, Gwynedd, 126, 130, 133
barley, 44, 131, 138, 225, 230
barn owls, 229
BASF, 104
bats, 7
beans, 22, 24, 25, 27, 41, 103, 107–8, 131, 231
Beauty of Bath apples, 21
beavers, 212, 215
bed and breakfast, 221
bees, 7, 18, 22, 25, 133, 140, 224, 255
beetroot, 22
behavioural change, 263–81, 288–90
    biofuels, reduction of, 289–90
    meat/dairy, reduction of, 264–7, 269–75, 288–9, 292
    overeating, reduction of, 263, 268–9, 277–8, 280
    waste, reduction of, 265–6, 267–9, 275–7
Belgium, 164, 167
Benin, 285

Benton, Tim, 278–81
'Best and Ward' study (1956), 35–6
Big Agriculture, 144–6, 235–7,
    240–41
bilberries, 139
Bill & Melinda Gates Foundation,
    232–3
bio-stimulants, 144
BioCarbon Registry, 196
biochar, 165–8
biodiesel, 78, 79, 96
biodiversity, 1, 4, 6–9, 12, 14, 207,
    239–40
    arable farming and, 114–15
    biomass and, 92
    endangered species, 10, 14
    fertilisers and, 46
    food production and, 7, 212, 225,
        239–40
    herbicides and, 46
    livestock farming and, 134, 263
    peat and, 153
    pesticides and, 46
    rewilding, 6, 47, 58–9, 125, 202–3,
        211–29
    solar farms and, 48–59
    woodlands and, 185
biodiversity credits, 6–7, 161, 228
biodiversity net gain (BNG), 226–8
bioenergy, 4, 43, 78–9, 88–101, 256,
    266, 289–90
    biomass, 79, 88–93, 190, 191, 256
    carbon storage and, 256
    liquid fuels, 43, 78, 79, 93–101
    peat and, 160
    reduction of, 266
    space efficiency, 98–9
BioEnergy with Carbon Capture and
    Storage, 92–3
bioethanol, 78, 79, 93–101
biomass, 79, 88–93, 190, 191, 251–2,
    256
birch trees, 129, 185, 201
birds, 18, 19, 50, 52, 115, 140, 149,
    222–3
Birmingham, West Midlands, 33–4
blackcurrants, 22, 248
Bluetop Solar Parking, 72–3

boar, 212
bogs, 125, 128, 150, 170, 235
    see also peatlands
Bolivia, 286
boreal forests, 209
Bosch, Carl, 104, 105, 260
Botley West, Oxfordshire, 44, 46
Bottle, Clare, 64–72
bramblings, 223
Bramley apples, 21, 175
Branson, Richard, 94
Brazil, 175, 193, 240, 286, 287
break crops, 247, 252, 258
Brewer, Andrew, 56–9
Brewood Park Farm, Staffordshire,
    114–22, 222–6
Brexit, 66, 279
Bridle, Sarah, 156, 264–6
Brigadoon (musical), 177
Bristol, England, 87
British Broadcasting Corporation
    (BBC), 199, 226
British Farming Awards, 56
broadleaf docks, 50
broccoli, 161
Bronze Age, 238
Broughton Estate, North Yorkshire,
    211–22
brown rust, 117
Brown, James, 161–8
Browning, Helen, 266, 272–4, 276
BTG Pactual, 193
bullrush, 170
Bunloit, Highlands, 211
business world, 6, 10
    carbon credits and, 199
    solar power and, 63–71
    wind energy, 87
buttercups, 218
butterflies, 48, 203, 215, 229
buzzards, 203

Caledonian pine, 184, 290
California, United States, 167
calories, 27, 41, 264–5, 266, 268
Cambridge Crop Science Centre, 234
Cameron, David, 20–21
Campaign for Nuclear Disarmament, 44

camping, 221
Canada, 175, 193, 209
cannabidiol oil (CBD), 241–2, 246
cannabis, 241–2, 244, 248
Caplor Energy, 63
car parks, 72–6
Carbon Capture and Storage (CCS),
    92–3
carbon capture, 111, 189, 225, 242–9
    hemp and, 242–8
    perennial crops and, 248–9, 257
    waste product fertiliser and, 262
carbon credits, 6, 161, 173, 194–9
carbon dioxide (CO$_2$) emissions, 1, 2,
    3, 5, 7, 15, 153
    biomass and, 92–3, 190, 191
    cement and, 189
    deforestation and, 14, 152, 285–7
    degradation and, 291
    fertilisers and, 105, 259–62
    food waste and, 268
    livestock farming and, 14–15, 133,
        139, 251
    peat and, 14, 151, 152, 153, 159,
        160, 163, 172, 173, 291
    scopes, 261
    steel and, 189
    timber and, 189–94
Carbon Positive Motorsport, 196
carbon storage, 1, 10, 242–9, 256–62
    biomass and, 92–3, 190, 191
    cover crops and, 225, 255, 257
    gardens, 23–42
    hemp and, 242–8, 257
    livestock farming and, 131–3, 137,
        251–2, 258–9
    pasture and, 14–15, 131–3, 137,
        139, 251
    peat, 14, 150–53, 158–9, 160,
        163–8, 172–3, 291
    perennial crops and, 248–9, 257
    pyrolysis, 165–8
    rewilding and, 225
    solar power and, 46, 56, 57
    trees, 3, 14, 159, 188–99
    waste product fertiliser and, 262
    windrow composting and, 111
Carbon Trust, 262

Cargill, 261
carnivorous plants, 235
Carson, Rachel, 5
Castro, Fidel, 39
cattle, 5, 83–4, 110, 127–8, 137–40,
    161, 200
    rewilding and, 217
cauliflowers, 161
CCm Technologies, 259–62
celery, 159
cement, 189
Centre for Ecology and Hydrology,
    158, 163, 167–8
Centre for High Carbon Capture
    Cropping, 242
cereals
    desert origins, 159
    fungi and, 103, 117, 233, 234–6,
        238–9
    nitrogen fixation and, 103–4,
        230–33
chaffinches, 223
Chalara fraxinea, 204
Changhua, Taiwan, 84
Charlotte potatoes, 22
Chatham House, 278
chemical fertilisers, 104–7, 111, 115,
    117, 121, 131, 143–7, 161, 259
    companies, influence of, 144–6,
        235–7
    foundational trials, 142
    'limited toolbox' problem, 146
Chernobyl disaster (1986), 80
cherry trees, 129, 196, 201
Cherwell Collective, 30–32, 47–8
Cheshire Wildlife Trust, 224
chickweed, 218
Chile, 104
chillies, 22, 24
China, 12, 45, 77, 243–4, 277, 286
chocolate, 261
Christmas trees, 201
Circular Bioeconomy Alliance, 192
Clarke, Richard, 196–8
class, 29–30, 37
clear felling, 177–8, 182, 187
Climate Change Committee, UK, 6,
    44, 46, 92, 154, 267

climate crisis, 2, 5, 12, 14–15
  arable farming and, 112–13
  biodiversity and, 208
  bioenergy and, 92, 93, 94, 96–7,
    100–101, 160
  droughts and, 148–9
  fertilisers and, 121, 230, 233
  food waste and, 268
  homes and gardens and, 30, 32
  hydropower and, 83
  imports and, 156
  livestock farming and, 127, 137,
    251, 253, 264–7
  nuclear energy and, 80
  peatlands and, 152, 171, 291
  solar energy and, 44–6, 49, 60, 68,
    71, 76, 83
  timber and, 189–94
  wind energy and, 88
clothing, 1, 243–5
clover, 48, 103, 131, 132, 231, 253
coal, 79, 100, 160
cocksfoot, 131
cocoa shells, 261
Common Agricultural Policy, EU, 6,
  200
common ragwort, 50
Community Power, 87
compost, 23, 31
conifer trees, 125, 140, 174, 177, 196,
  216
Connally, Emily, 30–32, 47–8
Conservative Party, 219
Constable, John, 44
Continuous Cover Forestry (CCF),
  177–88
Corby, Northamptonshire, 63–4
Corfield, Niels, 141, 143, 146–9
Cornwall, England, 56–9
Corsican pine, 201
Costa Rica, 285, 286
Costing the Earth, 93
cotton, 1, 244
Countryfile, 85, 199
courgettes, 22
cover crops, 108–9, 115, 118, 119,
  153–4, 225, 243, 255–6, 257
crab-apple trees, 129

cranesbill, 50
Creacombe Solar Farm, Devon, 48–52
credits
  biodiversity credits, 6–7, 161, 228
  carbon credits, 6, 161, 173, 194–9
cricket bats, 90
crop diversity, 258
croplands, 7, 13, 97, 102–22, 131, 138,
  235
  break crops, 247, 252, 258
  cover crops, 108–9, 115, 118, 119,
    153–4, 225, 243, 255–6, 257
  fertilisers, see fertilisers
  perennials, 248–55, 257
  protection of, 287–8
  soil, see soil
Cuba, 39
cuckoo flower, 170
Czech Republic, 151

Daily Mail, 280
Daventry International Rail Freight
  Terminal, 65
Davey, Edward, 98, 289
deep-rooted grasses, 148–9
deer, 125, 185, 203, 206
deforestation, 4, 9, 13, 14, 15, 95,
  285–7
DEFRA, 235–6, 243
degraded land, 5, 16, 138, 141, 167,
  172, 197–8, 290
Democratic Republic of Congo, 83,
  286
Derbyshire, England, 198
desalination plants, 155
deserts, 159, 291
diesel, 78, 79, 95
diet, 263–81
  food waste, 11, 16, 263, 265–6,
    267–8, 275–7
  meat/dairy reduction, 264–7,
    269–75, 288–9, 292
  overeating, 263, 268–9, 277–8, 280
'Dig for Victory' campaign (1940–45),
  35, 278–9
digging, limiting of, 23–4, 108–11,
  116, 118
Dimbleby Report (2020–21), 267

Diouf, Jacques, 99
Distributed Network Operators (DNOs), 66–7
Dobie, Ellinor, 174, 177–88
dock, 50
dog rose, 218
Domiz refugee camp, Kurdistan, 37–9
Doncaster, South Yorkshire, 65, 160, 166
Douglas fir, 178, 185, 196
doves, 203
dragonflies, 203
Driver, Alastair, 211–22, 225, 228
droughts, 148–9
Dubai, UAE, 72
ducks, 137
Duxford, Cambridgeshire, 239

Eat Local, 156
ecosystem services, 135
ecotourism, 285
Eden Renewables, 53–4
edge effect, 203
edible maps, 32–9
electrolysis, 99
elephant moths, 218
Elizabeth I, Queen of England, 244
elk, 212
Ely, Cambridgeshire, 157–60
endangered species, 10, 14
energy, 1, 4, 10, 11, 43–77, 78–101, 271
    bioenergy, *see* bioenergy
    fossil fuels, 1, 2, 9, 79, 80–81, 96, 100, 151, 155, 290
    hydropower, 77, 78, 79, 80, 82–3
    nuclear, 78, 79, 80
    solar, 1, 4, 10, 20, 21, 43–77, 81
    wind, 4, 10, 20–21, 57, 84–8
England
    biodiversity credits in, 6–7, 161, 228
    House of Lords report (2023), 3
    peat in, 150, 151
    solar energy in, 70
    wind energy in, 88
    see also United Kingdom
Ennerdale, Lake District, 211
Environment Agency, 110, 170, 213

Ethiopia, 82
Europe, 15
    bioenergy in, 95, 96
    cannabidiol oil in, 246
    obesity and, 268
    solar energy in, 60
    tree cover in, 15, 198
    wind energy in, 88
    woodlands in, 176–7, 189
European Union (EU), 278–9
    bioenergy in, 95, 96
    Brexit, 66, 279
    Common Agricultural Policy, 6, 200
    Emissions Trading Scheme (ETS), 173
    hemp in, 246
    Renewable Energy Directive (2018), 95
    solar energy in, 60
    wind energy in, 88
    woodlands in, 176–7
Evans, Chris, 158–60, 163
extinctions, 207

fake meats, 270, 271, 272, 273, 292
fallow deer, 203, 206
Far from the Madding Crowd (Hardy), 253
Farmers Weekly, 117
Farmers' Union of Wales, 136
farming advisors, 141
farming, 1, 4, 5–16
    arable, 102–22
    climate crisis and, 112–13
    dangers of, 134–5
    fertilisers, *see* fertilisers
    indoor farming, 32, 155, 165–6
    land crunch, 3, 282
    livestock, 123–41, 264–75
    mixed farming, 113–14, 148, 161, 200
    organic farming, 143–4, 161, 237–9
    protection of, 287–8
    regenerative farming, 114–22, 141, 144–9, 237–9, 283–4, 292
    soil, *see* soil
    urban farming, 24–42
    wildlife and, 111–12, 115

Farming and Wildlife Advisory Group, 6

Farrington, Duncan, 226

fat hen, 218

Feed in Tariff (FiT), 60–62

Fenland Soil, 153

Fens, England, 150, 151–60

fertilisers, 5, 27–8, 46, 95, 101, 103–11, 115, 230–39, 259–62
    chemical, *see* chemical fertilisers
    companies, influence of, 144–6, 235–7, 240–41
    cover crops, 108–9, 115, 118
    foliar feeds, 28, 120–21
    foundational trials, 142
    fungi and, 104, 234–6, 238–9
    greenhouse gases and, 105, 259–62
    guano, 104
    kelp, 104
    kitchen waste, 23, 28, 260
    nightsoil, 28, 38–9, 260
    nitrogen fixation, 103–4, 230–37
    waste of, 105
    waste products, 23, 28, 259–62
    windrow composting, 109–11

fibre crops, 243

fibrous material, 260

field maple, 201

finches, 223

Fine English Cottons, 244

Finland, 86, 198

fires, 2, 172, 174–5, 192, 194

firewood, 88–93, 137, 190, 191

First World War (1914–18), 89

fishing, 208

Flatford Mill (Constable), 44

floating farms, 83–4

floating solar, 77, 82–4

floods, 2, 10, 148

food; food production, 1, 2, 3, 5, 7, 10, 11, 15, 136–7, 263–81
    allotments, 17–20, 22–3, 35–6
    biodiversity and, 7, 212, 225, 239–40
    carbon storage and, 257–9
    cereal nitrogen fixing and, 230–37
    gardens, 21–2, 24–42
    greenhouse growing, 156

home delivery, 69
    importing of, 154–7, 164, 173, 225, 240
    intensive farming, 5, 9, 114, 135, 224–5, 243, 262, 269, 272, 283
    livestock farming, 127, 129, 131, 134, 136–7
    local sourcing, 156
    nationalism and, 279–80
    paludiculture, 170–71
    perennial crops, 248–55, 257
    population growth and, 5, 7, 58–9
    price of, 275–6
    rewilding and, 212, 225
    solar power and, 44, 46, 56, 58–9, 69
    subsidisation of, 168
    urban farming, 24–42
    variety of, 5, 258
    waste of, 11, 16, 263, 265–6, 267–8, 275–7

Food and Climate Change (Bridle), 156, 264

forest gardening, 31

Forest Research, 78

forests; woodlands, 174–206, 290
    afforestation, 6, 39, 188, 192, 194–206, 290
    bioenergy and, 88–93, 95
    carbon storage, 3, 14, 159, 188–99
    clear felling, 177–8, 182, 187
    commercial forestry, 10
    Continuous Cover Forestry (CCF), 177–88
    deforestation, 4, 9, 13, 14, 15, 95, 285–7
    fungal diseases, 204–5
    meat consumption and, 267
    monocultures, 177, 197
    rewilding and, 215, 217–20, 227
    sustainable harvesting, 190
    timber, *see* timber
    wildfires, 174–5, 192, 194

Fort William, Highland, 124

fossil fuels, 1, 2, 9, 79, 80–81, 96, 100, 151, 155, 290

fracking, 80–81

France, 74, 275

Friends of the Earth, 241

frogs, 19
fungi, 24, 103, 115–18, 180, 233, 234–6, 238–9
fungicides, 5, 115, 146

G's Fresh, 151–7, 159
Gamble, Charles, 86–7
Gandhi, Mohandas, 260
garden waste, 28
gardens, 17–42
    carbon storage, 23–42
    energy, 20–21
    food production, 21–2, 24–42
    recreation, 18
    urban farming, 24–42
    wildlife, 18–20
Garnett, Tara, 269–71, 275
gas, 1, 79, 80–81, 95, 96, 100
geese, 137
genetic engineering, 234, 238, 239, 240
Germany, 35, 41, 72, 76, 170, 211, 227
Ghana, 286
glamping, 221
Glenfeshie, Highland, 125
Glenfinnan, Highland, 124
Global Biodiversity Framework, 207–8, 239–40
'Global Biomass of Wild Mammals, The' (Greenspoon et al.), 208
Global Food Security Programme, 278
Global Forest Watch, 285
globe artichokes, 22
glyphosate, 118
Gold Standard, 196
golden plovers, 128, 140
goldfinches, 203, 223
Gove, Michael, 136
Grand Barry, France, 211
grass snakes, 20, 203
grasses
    bioenergy from, 78, 79, 100
    deep-rooted grasses, 148–9
    miscanthus, 78, 79, 243, 258
    monocultures, 148, 252
    multi-species, 49, 134, 138, 241, 252, 258
    ryegrass pastures, 130, 131, 252
    solar farms and, 49, 50, 56, 57

grasshoppers, 215
Grassland Farmer of the Year, 56
Great Fen project, 170
Green Alliance, 2
Green Business Watch, 61
green energy, see under energy
green hay, 49
green manures, 47, 108–9
Green Party, 219
Green Revolution, 105
greenfinches, 222
greenhouse gases, 5, 15, 95, 105
    carbon dioxide, see carbon dioxide
    fertilisers and, 105, 259–62
    food waste and, 268
    methane, 5, 99, 105, 133, 251, 253
    nitrogen oxide, 95
    nitrous oxide, 5, 105, 262
    peat and, 153
greenhouses, 156
Gregerson, Ole, 72–3
grey partridge, 223
grey squirrels, 205–6
grey water, 38
Gringley on the Hill, Nottinghamshire, 160
guano, 104
Guinness Book of Records, 102

Haber, Fritz, 104, 105, 260
haemoglobin, 231–2
halo thinning, 178
Hammond, Peter, 259
Hardy, Thomas, 253
Harry Potter Viaduct, 124
Havana, Cuba, 39
Hawes Water, Lake District, 211
hawthorn trees, 129, 217, 218
hazel trees, 201, 218
Health and Safety Executive (HSE), 134–5
health; health systems, 4, 280
Heart of England Forest, 198
heavy goods vehicles (HGV), 69
hedgerows, 135, 223–4
hemp, 48, 241–8, 257, 258
hemp agrimony, 170

hempcrete, 245
herbal leys, 252, 258
herbicides, 5, 46, 118
herbs, 24, 26
heritage of crops, 159
heritage varieties, 238–9
herons, 137
Hewitt Studios, 75–6
Highland Carbon, 196
Highlander (1986 film), 124
Highlands, Scotland, 123–6, 172, 196–7, 290
ten Hoeve, Paul, 222
holiday cottages, 221
Holme posts, 152
home food delivery, 69
homes; housing, 10
    carbon storage, 23–42
    energy production, 20–21, 59–63
    food production, 21–2, 24–42
    recreation, 18
    urban farming, 24–42
    wildlife and, 18–20
    wind turbines and, 86
hornbeam trees, 185, 201
House of Lords, 3
hoverflies, 224
Howell, Phil, 236, 250
Hull, East Yorkshire, 41
Humber Head Levels, 160–68
Humberside, England, 93
hunting, 14, 208
hydrogen, 99, 105
hydropower, 77, 78, 79, 80, 82–3
hypothecation, 271

Idle River, 164
imported food, 154–7, 164, 173, 225, 240
India, 12, 13, 72, 90, 277
Indonesia, 170, 172, 175, 286
indoor farming, 32–3, 155, 165–6
industrial zones
    solar energy, 63–71
    wind energy, 87
Innovation Farmer of the Year, 122
insecticides, 115, 223–4
insects, 19, 51, 129, 149, 223–4, 228

Institute for Agriculture and Horticulture, 293
Institute of Grassland and Environmental Research, 131
intelligent farming, 122
intensive farming, 5, 9
    arable farming, 114, 224–5, 243, 262, 283
    livestock farming, 135, 269, 272, 283
Intergovernmental Panel on Climate Change, 15, 284–5
International Energy Agency, 77
International Union for the Conservation of Nature, 221
IPBES, 4, 14
Iraq, 37–9
Ireland, 172, 198
irrigation, 148, 169
Isle of Mull, 123, 195–6

jacksnipe, 223
Jerusalem artichokes, 27
JoJu Solar, 72
Journal for Global Environmental Change, 287
Jurassic Park (1993 film), 124

kelp, 104
Ken Hill, Norfolk, 211
Kendall, Peter, 94, 96–7, 100
Kernza, 249–51
kestrels, 203, 223, 229
kingfishers, 170, 203
Kisielewski, Pawel, 259
kitchen waste, 23, 28, 260
Knepp Estate, Sussex, 211
Komorebi, 75
Krzywoszynska, Anna, 143

Labour Party, 219
lacewings, 224
Lake District, England, 211
Lancashire, England, 85, 244
Lancaster University, 54
'land crunch', 3, 282
Lapwing Estate, Humber Head Levels, 160–68
lapwings, 223

larch, 184
leaf litter, 260
leather, 1
legumes, 103, 230–33, 252–3, 255
Leicestershire, England, 198, 260
Leipzig University, 76
leisure, 4, 10, 26
Lemon Tree Trust, 37–9
lentils, 231
lettuce, 156
Liberal Democrats, 219
von Liebig, Justus, 143
lime trees, 201
Lindisfarne, Northumberland, 103
Ling, Richard, 261
Linking Environment and Farming
    (LEAF), 6
linseed, 112
liquid fuels, 43, 78, 79, 93–101
Little Carr Farm, Nottinghamshire, 164
livestock farming, 5, 9, 15, 16,
    123–41, 264–75
  behavioural change and, 264–75
  carbon/emissions storage and,
    131–3, 137, 251–2, 258–9, 264
  intensive, 135, 269, 272, 283
  mixed farming, 113–14, 148, 161,
    200
  rewilding and, 216–17
  sainfoin and, 252–5
  solar energy and, 47, 56, 57
  wildlife and, 133, 137, 140, 216–17,
    263
  wind energy and, 86
livestock on leftovers approach, 270
Loch Ness, Highlands, 211
Loch Shiel, Highland, 123–4
London, England, 19, 21, 22, 24–5,
    29, 36, 76, 84
Lopes, Harry, 53–4
Lost World (Doyle), 177
Lough Neagh, Northern Ireland, 106
low-input farming, 121
Lula da Silva, Luiz Inácio, 286
lynx, 212

Macalpine, William, 90–91
MacKay, David, 264

Mackie, Lindsay, 63
Madeley-Davies family, 126–41, 252
Magna Park, Lutterworth, 65, 67
maize, 41, 99, 103, 131, 157, 160, 230
'Making the most out of England's
    land' (2023 report), 3
malaria, 169
Malaysia, 167, 286
Mammal Society, 223
mammoths, 209
Manchester Moss, 160
mangroves, 290
manna grass, 170
maps, 32–9
Mark Wishnie, 193
Marks & Spencer, 126
Marshall, John, 195
marshes, 3, 150–51, 169
Matrix, The (1999 film), 142
meadows, 48–52
meadowsweet, 170
meanwhile gardens, 33
meat, 11, 16, 263, 264–7, 269–75,
    280–81, 288–9, 292
mental health, 4, 116, 284
Merrick, Sarah, 86
Merseyside, England, 93
Mesopotamia, 159
methane, 5, 99, 133, 251, 253
Mexico, 39–40
mice, 140
Microgeneration Certification Scheme
    (MCS), 62–3
miscanthus, 78, 79, 243, 258
mixed farming, 113–14, 148, 161, 200
Monbiot, George, 211, 216
monocultures, 113, 131, 148, 177, 197,
    200, 252
Monroe, Wisconsin, 96
Morpeth, Northumberland, 74
mosaic habitat, 203
mosquitos, 169
moths, 115, 169, 218, 222
multi-species grasslands, 49, 134, 138,
    241, 252, 258
'Multifunctional Landscapes' (2023
    report), 3
muntjac deer, 203

mushrooms, 103
Musk, Elon, 72–3
mustard, 108
mycorrhizal fungi, 103, 180, 233, 234–6, 238–9

National Diet and Nutrition Survey, 275
National Farmers' Union, 94, 96
National Food Strategy, 267
National Forest, 198
National Grid, 135
National Highways, 199
National Institute of Agricultural Botany, 235–7, 239, 243, 251, 255
National Parks, UK, 85, 135
nationalism, 278–9
Natural England, 110, 170
Natural History Museum, 210
Natural Resources Wales, 127
Nature Conservancy, 176
Nature Friendly Farming Network, 5–6
Nature, 81
NatureScot, 187
nematode worms, 254
Nestlé, 260
Net Sink, 286
net zero goals, 2, 6, 9, 46, 92, 262
Net Zero Review, 46
Netherlands, 83–4, 150, 151, 164, 167, 198
nettle tea, 19, 28
New Zealand model, 161
Newcastle, Tyne and Wear, 26
newts, 19, 203
Next, 64–5
nightsoil, 28, 38–9, 260
NIN, 232
nitrate vulnerable zones, 106
nitrogen; nitrate fertilisers, 46, 48, 104, 107–8, 111, 116, 120–21, 131, 142, 146–7, 231–7
   fungi and, 103, 180, 233, 234–6
   legumes and, 103, 230–31, 233
   sainfoin and, 255
nitrogen oxide, 95

nitrous oxide, 5, 105
Nobel Prize, 230, 233
North America, 4, 7, 15, 190, 193, 268
   see also Canada; United States
North Sea, 112
Northern Ireland, 88, 106
Northumberland, England, 74, 102–3
Norway spruce, 184, 185
Norway, 86
NSP2 gene, 239
nuclear power, 78, 79, 80

oak trees, 129, 134, 137, 185, 196, 201, 205–6
oats, 138, 225
obesity, 11, 268–9, 277–8, 280
Octopus, 70
Oder river delta, 211
OFGEM, 67
oil, 1, 96, 100
oil palm plantations, 95, 171, 172, 248, 286
oilseed rape, 44, 78, 97, 222, 225, 226, 258
Oldroyd, Giles, 230–41
onions, 22
orangutans, 96
Oregon State University, 57
Organic Farmers and Growers, 6
organic farming, 10, 143–4, 161, 237–9
ospreys, 203
Our World in Data, 5, 84, 155, 263, 277
overeating, 263, 268–9, 277–8, 280
OVO, 199
owls, 140, 229
ox-eye daisies, 112

Pakistan, 90
Palahi, Mark, 192
palm oil, 95, 171, 172, 248, 286
paludiculture, 170–71
Parker, Guy, 48–9
parking lots, see car parks
Parques Huerta, 40
Parton, Mackenzie, 223
Parton, Tim, 114–22, 145, 222–6
partridge, 223

pastures, 3, 7, 13, 14–15, 130–41, 283
  as 'bonsai', 148
  carbon storage and, 14–15, 131–3,
    137, 139, 251
  ryegrass, 130, 131, 252
  wildlife and, 133, 137, 140
peanuts, 231
pear trees, 201
peas, 22, 103, 138, 161, 231
Peatland Code, 196
peatlands, 14, 150–73
  carbon storage, 14, 150–53,
    158–60, 163–8, 172–3, 291
  Ely conference (2023), 157–60
  erosion of, 152–3, 158, 163
  meat consumption and, 267
  paludiculture, 170–71
  pyrolysis, 165–8
  rewetting, 164, 167, 168, 170, 172
  solar energy and, 168, 169
  water table and, 159, 164
PepsiCo, 260
perennial crops, 248–55, 257
Peru, 13, 104
pesticides, 5, 46, 161, 236–7
phacelia, 108
pheasants, 203
phosphorus, 104, 111, 116, 120, 147,
  235
photosynthesis, 31, 43, 94, 100, 103
Pink Fir Apple potatoes, 22
pipits, 140
plankton, 207
plastic pollution, 209
plough pan, 108
ploughing, limiting of, 23, 108–11,
  116, 118
Poland, 151, 211
Pollan, Michael, 290
pollinators, 27, 33, 52, 112, 218, 224,
  255
pollution, 2, 6, 9, 95, 208, 209
  air, 2, 6, 9, 40, 80, 95, 209
  water, 9, 91, 105–6, 209
polytunnels, 38
ponds, 19, 137
poplar trees, 201, 204–5, 243
poppies, 162–3

population growth, 5, 7, 58–9
Portishead, Somerset, 76
Portugal, 82, 155
potassium, 104, 111, 116, 147
potatoes, 22, 24, 27, 41, 131, 258
Proceedings of the National Academy
  of Sciences, 208
processed food, 37, 269
Project Drawdown, 249
protein, 27, 36, 41, 265, 270, 273
  alternative proteins, 270, 271, 272,
    273, 292
  animal feed, 132, 138, 140, 253
public footpaths, 130
purple emperor butterflies, 203
Purvis, Andy, 210–11
pyrolysis, 165–8

RAC Foundation, 74
Radio Four, 226
radish, 108, 155
ragwort, 50, 218
railways tracks, 77
rainforests, 175–6
rallying, 194–6
Ramage, Michael, 191–2
Random Acts of Greenness, 29
Ranganathan, Janet, 283, 286, 289
Rannoch Moor, Scotland, 125
rapeseed, 44, 78, 97, 131, 222, 225,
  226, 258
raspberries, 22, 130, 248
recreation, 4, 10, 26
reed, 170
reed buntings, 223
refugees, 37–9
regenerative farming, 114–22, 141,
  144–9, 237–9, 283–4, 292
Renewable Energy Directive (EU,
  2018), 95
resilience, 179
restoration, 290–92
Retford, Nottinghamshire, 166
reverse coal, 160, 165–6
rewilding, 6, 47, 58–9, 125, 202–3,
  211–29, 292
  spectrum approach, 211–12, 229
Rewilding Britain, 211–22

Rhine River, 151
rhizobia, 103
rice, 5, 41, 103, 230
Ripple Energy, 86
Ritchie, Hannah, 263, 265, 271
robins, 203
roe deer, 203, 206
rooftop gardens, 33
Rosario, Argentina, 40
rosebay willowherb, 218
rotational grazing, 283
Rothamsted, Hertfordshire, 89–91,
    222
Rothschild, Charles, 169
Rotterdam, Netherlands, 83–4
Royal Forestry Society, 205
Royal Institute of International
    Affairs, 3
Royal Society, 3
Royal Society for the Protection of
    Birds, 223
Russia, 175, 209, 279
rust, 205–6
Rwanda, 285
ryegrass, 130, 131, 252

Saab, 93–4
Saemangeum, South Korea, 84
Sahara Desert, 45, 209
sainfoin, 252–5
salad veg, 22, 24, 155
salicylic acid, 90
Save the Rainforest, 175–6
sawmills, 181, 182, 184
scabious, 129–30
scarlet pimpernel, 218
schools, 71–2
Schwarzenegger, Arnold, 273–4
Scotland, 85, 86, 88, 123–6, 172,
    177–88, 196–7, 290
Scotland's Rural College, 293
Scots pine, 184, 290
Scottish National Party, 219
Searchinger, Tim, 194
Second World War (1939–45), 35,
    278–9
Senegal, 151, 155
Seoul, South Korea, 41

Septoria, 116–17, 145
sewage, 28, 260
shale gas, 80–81
sharing vs sparing, 10
sheep, 5, 47, 56, 57, 80, 86, 123,
    127–8, 129, 131, 133, 139
    rewilding and, 215, 216
    sainfoin and, 253–5
shiv, 245–6
short rotation coppice, 91
short-eared owls, 140
Shropshire, John, 151–7, 159
Silent Spring (Carson), 5
Silk Road, 243–4
silver birch trees, 201
silvo-pasture, 283
Site of Special Scientific Interest, 126,
    186
Sitka spruce, 184, 185, 188, 196
Skipton, North Yorkshire, 216
Sky News, 199
skylarks, 222, 223
Slow Food Birmingham, 33–4
slugs, 19
Smith, Lydia, 241–59
Smith, Mark Ridsdill, 24–30
Smith, Rod, 102–3, 107–14
smoking, 274
snails, 19
snakes, 20, 203
snipe, 223
Snowdonia National Park, Wales, 126,
    128, 135
social class, 29–30, 37
soil, 5, 6, 9, 31, 48, 107–22, 142–8
    'additive mindset', 143–4, 146
    bacteria in, 28, 110, 116, 120,
        146–7, 166, 180
    carbon storage, 14–16, 23–4, 111
    compaction of, 108, 113–14,
        118–19, 148, 216–17, 247
    cover crops, 108–9, 115, 118, 255–6
    degradation, 5, 16, 138, 141, 167,
        172, 197–8, 290
    digging, limiting of, 23–4, 108–11,
        116, 118
    erosion, 99, 109, 112, 116–17, 136,
        148, 152, 158

fungi in, 103, 115–18, 180
hemp and, 247–8
Kernza and, 249–51
livestock farming and, 136
organic farming, 143–4, 237–9
sanfoin and, 255
sterilisation for trials, 142
water-holding ability, 148, 216–17
worms and, 24, 27, 112, 116, 226
Soil Association, 6, 117, 118, 241, 255, 266
Soil Farmer of the Year, 122
solar energy, 1, 4, 10, 20, 21, 43–77, 79, 81
  car park production, 72–6
  carbon storage and, 46, 56, 57
  floating production, 77, 82–4
  food production and, 44, 46, 56, 58–9
  household production, 20–21, 59–63
  industrial building production, 63–71
  peatlands and, 168, 169
  railways track production, 77
  school building production, 71–2
  single axis tracking, 53–4
  space efficiency, 81–2, 98
  wildlife and, 46–59
Solar Energy UK, 51
Solar for Schools, 71
Solar Park Impacts on Ecosystem Services, 54–5
Solar Sense, 72
Somerset Levels, 160
sorghum, 230
South Africa, 34
South America, 15, 193
South East Asia, 95
South Korea, 41
soy, 157, 231
soy oil, 96
Spain, 45, 82, 151, 155, 156
sparing, 10
sparrowhawks, 203
spectrum approach, 211–12, 229
Speth, Gus, 282
sphagnum moss, 150, 170
spiders, 115, 224
spinach, 22

spring onions, 155
squash, 22
squirrels, 205–6
St John's wort, 50
Staffordshire, England, 198
starlings, 18
State of Nature Report, 224
steel, 189
Stella Artois, 135
Stevenson, Frances, 45
stinging nettles, 19, 28, 106
Storm Arwen (2021), 183
storms, 2
subsidies, 6, 53, 62, 90, 101, 127, 135
subsoilers, 108
sugar beet, 79, 131, 160
sugar tax, 277–8
Sun-Ways, 77
sunflowers, 112
Sustainable Farmer of the Year, 122
Swadling, Janet, 293
Sweden, 156, 176
sweet manna grass, 170
sweetcorn, 22
Switzerland, 77
sycamore trees, 185, 217, 218
Syria, 37–9

Taiwan, 84
tannins, 254
Tappin, Trevor, 119
tarmac gardens, 33
taxation
  meat and, 271, 280–81
  solar power and, 60, 62, 70
  sugar and, 277–8
teasels, 50
Teesside, England, 93
Telegraph, 280
Tesla, 72
tetrahydrocannabinol (THC), 242, 244
Texas, United States, 87
Thiaw, Ibrahim, 2
Thinopyrum intermedium, 249–51
39 Ways to Save the Planet (Heap), 176, 189, 226, 259
thistles, 50

timber, 10, 89–90, 100, 137, 140, 175, 176, 177–94
  carbon emissions and, 189–94
  clear felling, 177–8, 182, 187
  Continuous Cover Forestry (CCF), 177–88
timothy, 131
tomatoes, 22, 24, 25, 27, 156
Tomkins, Mikey, 32–9
Torne river, 166
Tough Mudder, 214
tourism, 221, 285
tractor roll-overs, 134–5
Transport and Environment, 95
tree cover, 15, 174, 188, 198
trees, 3, 14, 175
  afforestation, 6, 39, 188, 192, 194–206, 290
  bioenergy and, 88–93, 95
  carbon storage, 3, 14, 159, 188–99
  deforestation, 4, 9, 13, 14, 15, 95, 285–7
  fungal diseases, 204–5
  livestock farming and, 129
  rewilding and, 215, 217–20, 227
  timber, see timber
  wind energy and, 86
Trico, 185
tropical regions
  forests, 9, 175–6, 192
  mangroves, 290
  peatlands, 169, 171–2
Trump, Donald, 86, 279

Uganda, 83
UK Forestry, 177
UK Warehousing Association (UKWA), 65, 69, 70
Ukraine, 7, 36, 80, 96, 279
United Kingdom, 2–3, 7–9
  anaerobic digestion in, 98–9
  arable farming in, 102–22
  biodiversity in, 6–7, 161, 228, 239–40
  Brexit, 66, 279
  carbon credits in, 194–9
  Climate Change Committee, 6, 44, 46, 92, 154, 267

environmental groups in, 5–6
food imports, 7
food waste in, 275
hemp in, 244, 246
livestock farming in, 123–41
National Food Strategy, 267
net zero goals, 2, 6
Net Zero Review, 46
obesity in, 277–8
paludiculture in, 170–71
rewilding in, 211–29
Second World War (1939–45), 35, 278–9
solar power in, 20, 21, 44–76
subsidies in, 6, 53, 62, 90, 101, 127, 135
tree cover, 15, 188, 198
wind energy in, 20–21, 85–8
woodlands in, 15, 174, 177–206
United Nations, 279
  Biodiversity Conference, 207–8
  Clean Development Mechanism, 196
  Convention to Combat Desertification, 2 291
  deforestation, reports on, 285
  Food and Agriculture Organization, 99, 268
  Intergovernmental Panel on Climate Change, 15, 284–5
United States
  bioethanol in, 94, 95, 96, 100
  cannabidiol oil in, 246
  diesel in, 95
  farming accidents in, 134
  gas in, 80
  local sourcing in, 156
  rewilding in, 211, 227
  solar energy in, 57, 59–60
  Trump administration (2017–21), 279
  wind energy in, 87
University of Cambridge, 191, 230, 293
University of Oxford, 269
University of York, 264
Urban Agriculture Programme, 40
urban farming, 24–42

VCS, 196
veganism, 264–7
vegetarianism, 270
vertical farming, 34, 41, 154, 161
Vertical Veg, 25
voles, 203
Volkswagen, 95

wading birds, 112, 164
wagtails, 203
Wales, 86, 88, 126–41
Walkers, 260
walnut trees, 201
warehousing, 63–71
wars, 279–80
Warwickshire, England, 198
waste product fertilisers, 23, 28,
    259–62
water mint, 170
water sports, 130
water table, 159
water voles, 203
watercress, 170
waterfowl, 203
Watson, Robert, 4, 14
We Do Solar, 76
wellingtonia, 201
wetlands, 4, 106, 112, 157, 159, 164,
    169–70
wheat, 41, 44, 78, 79, 103, 107, 108,
    121, 131, 157, 225, 258
    fertilisers and, 230, 235–6, 238
    heritage varieties, 238–9
White Rose Forest, 218
Whiteadder Water, Berwickshire, 177,
    183, 186
wicker, 89–90
wild celery, 170
wilderness, 209
wildfires, 2, 172, 174–5, 192, 194
wildflowers, 18–19, 48–52, 106,
    129–30
wildlife
    afforestation and, 203
    arable farming and, 111–12, 114–15
    gardens and, 18–20
    livestock farming and, 133, 137,
        140, 216–17

meat consumption and, 267
rewilding, 6, 47, 58–9, 125, 202–3,
    211–29
solar power and, 46–59
wetlands and, 163, 164, 169–70
woodlands and, 178, 185
Wildlife Trust, 170
Williams, Gareth, 63
willow, 58, 78, 89–93, 196, 201, 203,
    218, 243, 258
Wincanton Distribution Centre, Corby,
    63–4
wind energy, 4, 10, 20, 57, 84–8
    homes and, 86–7
    industrial building production, 87
    public support for, 88
    space efficiency, 84–5
    visual impact, 85–6
windrow composting, 109–11
Wire, The, 236
Without the Hot Air series, 264
Witter, Lucy, 224
wolves, 212
woodcock, 223
Woodland Carbon Code, 196
Woodwalton Fen, Cambridgeshire,
    169–70
wool, 1
Woolwich, London, 76
Worcestershire, England, 198
Worksop, Nottinghamshire, 166
World Database of Protected Areas,
    285
World Resources Institute (WRI), 13,
    97–8, 100, 189–92
    four pillars, 15–16, 282–92
World Trade Organization, 279
worms, 24, 27, 112, 116, 226
wrens, 203
Wrigley, Rebecca, 211
Wychwood Biodiversity, 48–52

yellowhammers, 50, 223
Yellowstone National Park, 211

Zestec, 70
Zinco, 41